Entropy and the Magic Flute

Entropy and the Magic Flute

Harold J. Morowitz

OXFORD UNIVERSITY PRESS
New York   Oxford

Oxford University Press

Oxford   New York
Athens   Auckland   Bangkok   Bombay
Calcutta   Cape Town   Dar es Salaam   Delhi
Florence   Hong Kong   Istanbul   Karachi
Kuala Lumpur   Madras   Madrid   Melbourne
Mexico City   Nairobi   Paris   Singapore
Taipei   Tokyo   Toronto

and associated companies in
Berlin   Ibadan

First published in 1993 by Oxford University Press, Inc.,
198 Madison Avenue, New York, NY 10016

First issued as an Oxford University Press paperback, 1996

Oxford is a registered trademark of Oxford University Press

Library of Congress Cataloging-in-Publication Data
Morowitz, Harold J.
Entropy and the magic flute /
Harold J. Morowitz.
p.   cm.   ISBN 0-19-508199-4
ISBN 0-19-511134-6 (Pbk.)
1. Science—Miscellanea.   I. Title.
Q173.M87   1993   500—dc20   92-27935

The chapters in this volume originally appeared as essays in *Hospital Practice*.
We gratefully acknowledge reprint permission from both *Hospital Practice* and
The Maclean Hunter Medical Communications Group Inc., its publisher.

10 9 8 7 6 5 4 3 2 1

Printed in the United States of America

This volume is dedicated to
the memory of my mother

ANNA LEVINE MOROWITZ

Only decades later have I come to appreciate
the depth of her influence
on my love of learning

Contents

IV

V

VI

VII

Introduction

Some years ago I was asked to contribute a monthly piece on science and society to the journal *Hospital Practice*. This began a long relationship which has led to my becoming an essayist, a happening that would have come as a surprise to some of my high school English teachers.

Being an essayist has changed my life, and I want to express my gratitude to those who make it possible. Each page that I read, each sight that I see, becomes potential input to cluster around some central thought that may one day emerge as part of a commentary on science, society, and the human condition. Because of the requirement to produce with regularity, I think I examine some things more fully and experience some events more acutely than I did previously. For example, I now travel without a camera because I want to see without distraction.

I've thought about the essay as a literary genre and have a few ideas. The essay is by its nature idiosyncratic. Dealing with a single subject in a rather few words requires a kind of trimming that must reflect strongly on the views of the author. There usually is insufficient space for a balanced point of view. Indeed, Michel de Montaigne, who invented this form in the late 1500s, chose the word *essai,* an endeavor or trial or test, thus stressing the personal. Parenthetically, I might add that when my first book of essays was published, a reviewer in *Nature* noted that as an essayist I didn't come up to Montaigne. Alas, he was right.

The essay has historically ranged over philosophy and politics and occasionally has been a vehicle for humor. Indeed, looking back over the years I recall the first essayist I ever read regularly:

Stephen Leacock. In retrospect, I do believe that a youth spent reading Leacock was fine training in expressing a sense of humor and avoiding the sin of taking oneself too seriously. In my favorite Leacock story, he reported his worst nightmare, lecturing to an audience of 600 people, all of whom fell asleep. He then noted that he woke up and found out it was true.

In the twentieth century, the essay has developed into a mode of communication between scientists and a broader public. One of the best known writers of this school was J. B. S. Haldane, whose essays are held up as a paradigm of explanatory prose. Another personal aside: when my second book of essays was published, the same magazine *Nature* (as previously mentioned) noted in the review that my scientific writing was not as good as Haldane's. Too true, but those *Nature* reviewers sure are tough on a guy. Another of the early scientific essayists was G. Evelyn Hutchinson, whose *Marginalia* remains a lasting example of elegance and erudition. Among contemporary scientific essayists, Lewis Thomas and Stephen Gould have had great impacts on a very substantial reading public.

As I look back over the years, another event stands out with some clarity. Some months after I had begun writing for *Hospital Practice,* I was in my study at home talking to a visiting colleague, Professor Robert Haynes of York University. Bob is a fellow scientist-humanist, and we were discussing the essay as a literary form. I told him about my regular writing chore and voiced the concern that I might run out of things to write about. With a twinkle in his eye he walked over to the bookshelves, pulled down the appropriate volume of the *Encyclopedia of Philosophy,* and opened to an article, "Nothing" by P. L. Heath. "Look," he said, "if Heath can write a substantially long entry on nothing, do you really think you're going to run out of things to write about?" We laughed heartily.

There was, of course, a profound truth in Bob's demonstration. As philosophers point out, there is only one kind of nothing, but an infinity of somethings. Given that we live in a world of somethings, there is a lot to write about. Material has not been a problem. The world of the last seventeen years has been so dynamic, knowledge has emerged so rapidly, and change has so characterized our lives that there has been a constant need to address new issues and examine old issues in 'a new light.' That is, of course, one job of the essayist.

Writing an individual article is one thing, but when one collects them together into a book, a higher law must be invoked to order them into related themes. And so with the help of a friendly editor I have gathered my thoughts into seven heaps, as the Buddhists would say, and briefly present the rationale of this scheme.

The first section deals with a number of physicists and works of physicists that have had a profound influence on the way I have looked at the universe. Although my principal academic world has been in the field of biology, I never stray very far from the principles of physics that illuminate my view of almost everything. This section is a tribute to a number of savants who have penetrated the world of nature most deeply.

The next group of writings deals with scientific ethics in classical and biblical times as well as in the contemporary world. The first examples have come from serving on committees dealing with these issues. The later pieces emerge from my own and others' experiences in the world of the real. This section is not a search for the elusive moral absolute, but an attempt to find the way in each case under discussion.

Section three extends some of the issues of medical ethics and related problems to the fields of addiction, nutrition, and parturition. A number of practices in present day United States are examined in a probing way to seek the rational and irrational in the paradigms. They search for that glimmer of good sense in enigmatic human behavior.

The next pile of essays is uncharacteristically personal. It deals with my family, my college roommate, and institutions where I have worked. But it does happen that through individual experiences one reaches for deeper truths and hopes they emerge.

The fifth group of writings have a common feature. They were written on the road, on planes, and in hotel rooms. I delight in traveling, and each new location somehow suggests new thoughts. Thus these pieces come from first drafts written in Kalakol, Nairobi, Mombassa, Yokohama, Kyoto, Hilo, and Palo Alto. I eagerly await the next trip and that chance happening that triggers new thoughts, new ideas.

The academic life, the professoriate, has a series of attitudes and feelings all of its own. From time to time I try to capture this ethos with some anecdotes from the world around me. I hope that this is done in a good-humored way, but occasionally the good humored moves to enigmatic to ironic to sardonic. *Mea culpa*. People who get into academics should expect no less from their colleagues.

I save for last biology, my thing. Like the field itself, this collection covers the ground from life's origins to global ecosystems. The events that led to these writings took place in the laboratory, on the phone, at meetings, and at home. They lead to no overwhelming conclusion, but a historical science like biology is too rich to encompass in a grand unified theory. We have to savor the wonder of each genus, each species, and ultimately each individual. Then again, life is a property of the right kind of universe.

Fairfax, Va.   H. J. M.
November 1992

I

1

Entropy and the Magic Flute

IN THE DYING LIGHT of a gray December afternoon my wife and I stood before the tombstone of the family Boltzmann in the Vienna Central Cemetery. At the center was a carved stone bust of the eminent nineteenth-century physicist Ludwig Boltzmann. On the left was the name of his wife, and on the right were the names of their three sons, the youngest of whom had died at Smolensk during World War II. Inscribed in stone above the bust was the formula $S = k \log W$. The symbol S represents entropy, k has become known as Boltzmann's constant, and W is the number of possible molecular configurations corresponding to a given observed state of a physical system. It is the formula that launched the science of statistical mechanics.

Our visit to the Boltzmann memorial had its beginnings many years back when I was indexing a book on thermodynamics in biology. As the pile of 3×5 cards with the name Boltzmann began to pile up, it seemed strange that I knew so little about the scientist whose name figured so prominently in the development of a scientific discipline that was close to my heart. A trip to the library revealed only one biography: *Ludwig Boltzmann, Mensch, Physiker, Philosoph* by Engelbert Broda. I had once met the author of that book at a meeting on the origin of life. Broda, a physical chemist on the faculty of the University of Vienna, had a strong interest in prebiotic energetics. He had pointed out to me that Boltzmann was the first scientist to note the negative entropic role of photosynthesis in ordering living matter, an idea taken up some 70 years later by physicist Erwin Schrödinger.

I perused the biography, noting that it was too bad that the life story of this extraordinary man was not available in English. As a consequence, few people in the United States are aware of his enormous influence on the turn-of-the-century physics that led to the atomic age. Few also are aware of Boltzmann's tragic death by his own hand, brought about, according to some contemporaries, by a depression caused by the failure of his colleagues Ostwald and Mach to accept his theories of the atom.

My family is involved in a small publishing company, and so I persuaded the powers that be to look into the possibility of doing an English edition of the Broda biography. Some correspondence established that a translation had been done but not published and that Broda was interested in editing, updating, and briefly adding to the original for publication in the United States. He also had in his possession a remarkable collection of photographs of the scientists with whom Boltzmann interacted, and these were to be added to the book.

The English edition was clearly a labor of love for Broda, and at the edited-manuscript stage, the galley-proof stage, and the page-proof stage, he constantly added and amended. This is the kind of thing that drives publishers mad, but the Boltzmann book had also become a labor of love for them, and they went along with this constant polishing as the book slowly progressed toward publication. Finally, the finished volume appeared in 1983, and Broda expressed his pleasure. A few months later he died quite suddenly. It was in our exchanges with this author that we learned of the Boltzmann memorial with its celebrated formula.

On our arrival in Vienna, inquiries did not immediately lead us to Boltzmann, as he is not well known to contemporary Viennese. Indeed, it is safe to say that he is unknown to almost everyone. We were directed to the Central Tourist Office in the Opera Ring, where I inquired after the statue of Boltzmann. A most helpful lady looked him up in an archival work. There were two Boltzmann statues: one in the arcade of the university, which was closed that day, and the second in the Central Cemetery several kilometers away. Her first reaction was, "Why don't you go see the Mozart statue? It's right nearby." When she realized that we were not going to be satisfied with the substitution, she carefully detailed the best tram route to the second gate of the Central Cemetery.

During the ride, I mused over the Boltzmann-Mozart theme. It reminded me of those wonderful and seemingly endless sophomoric conversations about who was the greater genius, Shakespeare or Newton. A few decades of experience have rendered such issues less pressing, and indeed less meaningful, for true genius seems to defy the criteria of evaluation with which such questions can be posed.

When we showed the gatekeeper the slip of paper on which the tourist office agent had written "Ludwig Boltzmann" and the location of the tomb, he immediately responded to the "Ludwig B" part of the message and began directing us toward the tomb of Ludwig van Beethoven, another local of some significance. Again we were being treated to a vignette on science and music. This one had a historical echo, since I recalled reading that Boltzmann had once attributed his science to the inspiration of the music of Beethoven and the writings of Schiller. We settled the confusion about which Ludwig and found our way to Boltzmann's last resting place. And so I stood there thinking of

- The Stefan-Boltzmann radiation law

- The Boltzmann constant

- Boltzmann's H theorem

- The Maxwell-Boltzmann distribution

- The Boltzmann entropy formulation

There were also thoughts of the man and his time, which saw the beginning of the revolution in physics.

While so engaged in thought, I must have been roused by the memory of some inspirational opening chords of Beethoven. It seemed a pity to leave here without paying homage to the other Ludwig. As we walked toward the musicians' section of the cemetery, we passed a stone reading:

ENGELBERT BRODA
1910–1983
EIN LEBEN FÜR DIE WISSENSCHAFT
UND FÜR DEN FRIEDEN

Indeed, Broda's career had combined a deep interest in science, humanism, and peace. His grave site is within 100 meters of his hero, Boltzmann, with whom he shared many goals.

We located the Beethoven memorial in a group of three monuments, with Mozart in the center, flanked by Beethoven and Schubert. Nearby were the graves of Brahms, Johann Strauss the Younger, and a number of other famous composers. The musicians' section of the cemetery is a clear reminder of the extraordinary position of Vienna in eighteenth- and nineteenth-century music. Mozart settled there in the 1780s. Beethoven came to study with him there in 1787. Haydn was both a friend of Mozart and a teacher of Beethoven. Schubert went to school in Vienna and then stayed to write music in the early 1800s. And so it went through the century.

As it happened, we had come to pay homage to a great scientist and ended up thinking about some great composers. Boltzmann would have understood the connection.

2

The Perversion of J. Willard Gibbs

IT WAS A QUIET SATURDAY AFTERNOON. I was browsing the library looking for help with a mathematical difficulty I was having in representing the rotation of molecules and their optical isomers. I chanced on a copy of *Vector Analysis* by J. Willard Gibbs and was suddenly relieved. First, I had learned my vector theory from Leigh Page, who had overlapped with Gibbs at Yale and taught with Gibbsian notation, which was therefore familiar to me. Second, I had the feeling that the authoritative and farseeing Gibbs would have included the material I was looking for.

And sure enough, on page 334 began a discussion of a dyadic, a mathematical operator that represents a rotation and is called a versor. The operation of turning is a "version," which should be a familiar enough term for those who remember that turning a baby in utero is generally called cephalic version or podalic version, depending on whether the head or feet are brought to present. What was somewhat startling was to read, a little further on:

> *Definition:* a transformation that replaces each figure by a symmetrical figure is called a *perversion* and the dyadic which gives the transformation is known as a *perversor.*

In other words, a perversor converts a figure into its mirror image.

This meaning of "perversion" was not totally strange in 1901, when *Vector Analysis* was published, as the *Oxford English Dictionary* presents us with

> 2. *a. Geom.* The formation of the perverse of a figure; the perverse itself.
> 1881 MAXWELL *Electr. & Magn.* II. 415 They are geometrically alike in all respects, except that one is the perversion of the other, like its image in a looking glass.

Nevertheless, Gibbs was such a Victorian figure that associating him with perversions in any form retains its shock value. After all, the *Encyclopaedia Britannica* notes:

> He remained a bachelor, living in his surviving sister's household. In his later years he was a tall, dignified gentleman, with a healthy stride and ruddy complexion, performing his share of household chores, approachable and kind (if unintelligible) to students.

I would like to protest the parenthetic smear by the encyclopedia writer. There is nothing at all unintelligible about Gibbs. Indeed, his writing has such lucidity that a number of us still return to his works to seek clarification of difficult issues. He is thought by many to be the greatest scientist yet produced by the United States, but his depth is in no way associated with lack of intelligibility.

In any case, these thoughts about Gibbs's perversions did serve as a reminder that we have just observed the 150th anniversary of the great savant's birth, on February 11, 1839. Yale, his alma mater, is staging a symposium this month to celebrate the event.

Gibbs's life had a quiet, uniform aspect. The only son of Josiah Willard Gibbs, Sr., professor of theology and sacred literature at

Yale, he spent his childhood, undergraduate years, and graduate years entirely in New Haven, a few blocks from his birthplace. After receiving his graduate degree, he traveled with his sisters in Europe for nearly three years, during which time he attended lectures by the great scientists and mathematicians of the Continent. He then returned to the family home, where he lived for the rest of his life with his sisters and brother-in-law, Addison Van Name, who was university librarian.

The total lack of perversion, scandal, or any kind of excitement in Gibbs's personal life has provided special problems for his biographers. Nevertheless, two rather notable life histories of this remarkable man have appeared. The first of these, published in 1942, 39 years after Gibbs's death, was written by Muriel Rukeyser, better known in most literary circles as a poet. Rukeyser added meaning and understanding to the life of Gibbs by presenting it in the framework of the intellectual history of nineteenth-century United States.

Rukeyser's book begins with a recounting of the *Amistad* mutiny in 1839. Off the coast of Cuba, 53 Mendi Africans being held as slaves managed to free themselves from chains, kill the captain and cook, take command of the boat, and sail for home. By day they traveled east; by night two of the remaining Spaniards took the wheel and sailed northwest. The ship finally ended up near Montauk Point, Long Island, where the U.S. Navy boarded and sailed to New Haven.

The case was brought before the courts, where the Africans, speaking a dialect of Mendi unknown to anyone in Connecticut, were unable to tell their story. Enter Josiah Willard Gibbs the Elder. Gibbs met with the captives and acquired a list of Mendi words, identified through sign language. He then took the stagecoach to Bridgeport, where he got the packet boat to New York. He went from ship to ship in the harbor and located on a British naval vessel two sailors who spoke both Mendi and English. He returned to New Haven with the interpreters, and thus began a new chapter in one of the most fascinating legal cases in U.S. history.

In the final hearing before the U.S. Supreme Court, the Africans were represented by the old and ailing John Quincy Adams, former President and then congressman from Massachusetts.

The government was represented by Frances Scott Key, then U.S. attorney. Adams prevailed, the slaves were freed, and they returned to their African home, accompanied, of course, by missionaries.

Thus began the poet's telling of the life of the younger J. Willard Gibbs. She ends with a discussion of three American masters: Melville, Whitman, and Gibbs. In between, we find William James, Henry Adams, Ralph Waldo Emerson, Henry James, Charles Peirce, and the full richness of the great flowering of Arts and Sciences in nineteenth-century America. If Rukeyser finds any perversion in Gibbs, it is that "imperative is his loneliness, the creative loneliness of the impelled spirit." He gave up the warmth of human contact for the hot glow of probing the laws of nature with fierce insight.

Lynde Phelps Wheeler was a physicist who had been a student of Gibbs in the 1890s. In 1944, he responded to a request to "prepare an authoritative biography which would contain not only the story of Professor Gibbs's life, but particularly illustrate the applications of his theory to the subsequent development of the physical sciences." And indeed, he succeeded in drafting a book that included both the life of his mentor and the explanation of his scientific work. Thus, without benefit of equations, Chapter V expounds thermodynamics, Chapter VII discusses mathematics, Chapter VIII deals with optics, and Chapter X discourses on statistical mechanics.

Between the chapters on science in the Wheeler book there emerges the same Gibbs we saw in the earlier biography, a man whose "mode of living was constantly shaped to the demands of those fundamental intellectual interests which constituted his real happiness."

I think most of us find it difficult to envision individuals as self-sufficient as Gibbs. Perhaps Immanuel Kant exhibited that same kind of inner life. Genius remains unfathomable, and for our description of Gibbs's genius, we turn to the words of his student Wheeler:

It has rarely been given to a man to penetrate so far in advance of his contemporaries twice in a lifetime. If the "Equilibrium of Heterogeneous Substances" was ahead of its time, so was the *Statistical*

Mechanics, and by about the same interval. . . . With their coming into their own these two works, so complementary in their nature, so penetrating in their insight into the inmost secrets of nature, and each set forth with meticulous logic, will endure as two of the greatest monuments of the human mind.

3

Les Atomes

OCCASIONALLY, I get really excited about a 70-year-old book, or perhaps a 120-year-old book, and then find colleagues giving me strange looks. One such "quaint and curious volume of forgotten lore" that impels me to exuberance is *Les Atomes* by Jean Perrin. The work, which should be honored as a classic, has fallen into disuse and obscurity. I'm in the process of trying to restore interest in this book for the enlightenment of historians of science and the edification of students of elementary chemistry, who are presented with the atomic hypothesis as if it were Holy Writ.

Around the beginning of the twentieth century, a great many scientists accepted atomism as a satisfying explanation of experiments on the nature of matter. There were, however, two powerful voices in the vanguard of the antiatomists—Wilhelm Ostwald in Leipzig and Ernst Mach in Vienna. The former, a devotee of thermodynamics, disliked mechanical doctrines and doubted the reality of atoms, regarding them as mere mental artifices. One notes in his position several unresolved problems that later emerged in the developing philosophy of science. Mach's view was in line with Ostwald's, though perhaps somewhat more extreme in rejecting atoms. Among the allies of Ostwald and Mach were such scientists as Pierre Duhem and George Helm.

The chief advocate of the atomic hypothesis was Ludwig Boltzmann. In 1906, the great physicist took his own life. It has been suggested that his state of depression was driven in part by the failure of many scientists, including Ostwald and Mach, to accept the kinetic molecular view of matter that he had championed. In any case, this matter of atomicity was serious business to the scientific community.

Jean Perrin (1870–1942) was educated at the École Normale Supérieure and followed the views of his mentor, Marcel Brillouin, a strong advocate of Boltzmann and opponent of Ostwald and Mach. From 1906 to 1909, Perrin carried out an extraordinary series of observations on suspensions of particles small enough to undergo Brownian motion. Using the statistical mechanics of Boltzmann and the theoretical results on Brownian motion by Einstein and Roman Smoluchowski, Perrin was able, with only a microscope and stopwatch, to arrive at values of Boltzmann's constant by several different ways. From this parameter he was able to derive Avogadro's number, the crucial enumeration of particles per mole necessary to establish atomicity.

In 1913, Perrin gathered all his work on atomic theory into the book *Les Atomes*. Listed are 16 different determinations of Avogadro's number, many of which he had personally performed. The work was a rhetorical masterpiece convincing almost everyone except Mach of the validity of the atomic hypothesis. The work went through many editions and translations, an English version appearing in 1920. In 1926, Perrin was awarded the Nobel Prize in Physics.

By 1926 the Perrin-Boltzmann message had been so successful that no one doubted the atomic hypothesis. The remarkable book that had put it all together went unread. Chemistry and physics texts now begin by stating that all matter is composed of atoms.

The prose has the ring of a doxology: Praise atoms from which all matter flows. Teachers have given up informing students how the atomic hypothesis was validated. They have even given up informing themselves. One of the truly great scientific books of this century gathers dust on library shelves and is missing from all libraries established after 1930.

My own knowledge of this Perrin masterpiece came from Alois Kovarik, the Yale professor who taught me atomic physics. Kovarik had worked in Paris in the 1920s and was very aware of Perrin's work. He probably knew Perrin. *Atoms* was on the list of supplementary readings for the course Physics 32. The details are lost, but I read the book.

Some years ago, views were being solicited on teaching techniques for introductory college chemistry. It occurred to me that the first semester could be well spent presenting the atomic hypothesis as an idea that began with Democritus in the fifth century B.C. and came to full fruition with Perrin's work in 1913. If students really understood how that hypothesis was validated, they would have a much deeper comprehension of science. The standard material of chemistry could then be treated in a somewhat condensed form. Perrin's book could serve as a text.

When I discussed this idea with a number of younger physicists and chemists, it appeared that few knew of Perrin and even fewer knew of his work on the ultimate validation of the atomic hypothesis. The book was, of course, long out of print.

Owing to my association with the CEO of Ox Bow Press—my wife—the English translation of the book has finally been reprinted. Mea culpa, I have a conflict of interest, but not to worry: This book will not make any money. A copy now sits proudly on my desk.

The beginning deals with the history of the atomic hypothesis from Democritus to Lord Rayleigh. It provides the background for thinking in terms of atoms and molecules. There follows the kinetic theory of moving atoms and molecules. Under the rubric "Molecular Agitation," the work deals with ideas of random thermal motion. There is a discussion of Brownian motion and the distribution of colloidal particles in a column of liquid. Brownian motion is also treated according to the Einstein-Smoluchowski theory and its experimental validation. Perrin deals with fluctuations and their treatment from the point of view of statistics. One chapter evaluates Avogadro's number vis-à-vis the Planck theory of blackbody radiation, and another analyzes atomicity from the determination of the charge on the electron. Finally, the then newly discovered phenomenon of radioactivity is used to count numbers of particles.

The last table in the book (see below) is one of the great scientific syntheses of all time. All of the independent measurements of Avogadro's number converge on a common value. The metaphysical requirement of multiple connectivity has long been one of the major features of the philosophy of science, and no construct demonstrates this more fully than the number of particles per mole. It is the bridge between the world of observation and the world of atoms.

Phenomena Observed	N (Avogadro's Number)/10^{22}
Viscosity of gases (kinetic theory)	62
Vertical distribution in dilute emulsions	68
Vertical distribution in concentration emulsions	60
Brownian movement	
Displacements	64
Rotations	65
Diffusion	69
Density fluctuation in concentrated emulsions	60
Critical opalescence	75
Blueness of the sky	65
Diffusion of light in argon	69
Blackbody spectrum	61
Charge as microscopic particles	61
Projected charges	62
Helium produced	66
Radioactivity	
Radium lost	64
Energy radiated	60

The currently accepted value of Avogadro's number is 60.22 × 10^{22}. Much of the sureness of modern chemical science emerges from the firm foundation established in *Les Atomes*. Now that the book is once more available, I think I shall again begin the campaign to teach introductory chemistry from this point of view.

4

Tricentennial

ONE AFTERNOON IN LATE NOVEMBER of last year I was reading for amusement the Preface of *Treatise on Analytical Dynamics* by L. A. Pars. This elegant compendium of classical dynamics begins, not surprisingly, with a recognition of Newton's *Principia,* noting that the first edition appeared in 1687. I was struck by the realization that it was the 300th anniversary of one of the greatest publications of all time, and yet the year had passed with, to my knowledge, no remembrances or festivities. The anniversary seemed totally unmarked, at least in the spheres in which I travel. It was hard to imagine. I toasted Sir Isaac with a glass of wine at dinner that night and resolved to mount some more fitting commemoration of the event, however belatedly.

I was about to leave on extended travel and had no immediate opportunity to pursue the matter further. During the trip I did meet up with a British engineering student from London. I asked him if he knew of any celebration of the 300th anniversary, and he looked at me blankly. He was in fact unaware that Newton is buried in Westminster Abbey, the first scientist to be so honored. Somehow the great savant is not all that revered in contemporary England. My traveling companion ventured the explanation that generation after generation of students struggling through calculus had never forgiven Newton for the invention of that branch of mathematics. I refused to accept this cynical judgment, and we went on to other matters.

The leisure and opportunity to look into the Newton anniversary did not become available for several weeks, until I was able to avail myself of the reference section of Nairobi's MacMillan Library. It was a Friday noon, and the strains of a muezzin chanting prayers over a loudspeaker system were booming forth from the mosque next door. Although the library collection did not contain a copy of the *Principia,* there was a photograph of the title page.

Every section of this page is rich with meaning for British history and the history of science. The title itself can be rendered as *Mathematical Principles of Natural Philosophy,* in itself a strong platonic assertion that natural philosophy is based on a mathematical foundation and can be understood in those terms. Here Newton was following Galileo both in establishing mathematical methods in physics and in identifying space and time as the primary variables of mechanics.

The next section identifies the author as Lucasian Professor of Mathematics at Trinity College, Cambridge, and Associate of the Royal Society. Each of these designations is significant, for college and society played important roles in the emerging Newtonian epic.

The following lines come as something of a surprise: the imprimatur of Samuel Pepys, President of the Royal Society. The word "imprimatur" has a strange ring because we associate it with the nihil obstat, the permission to print a book granted by a bishop of the Roman Catholic Church. That such permission should be required of the president of the Royal Society seems contrary to the spirit of the parliamentary age in which Newton lived. The usage is perhaps a subtle antipapist statement about

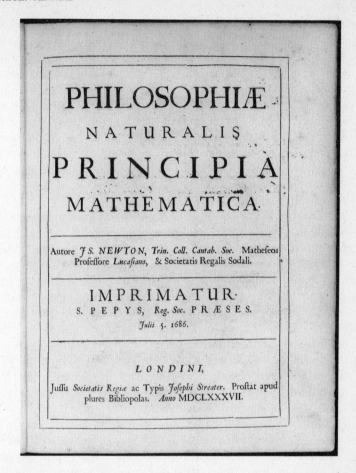

PHILOSOPHIÆ

NATURALIS

PRINCIPIA

MATHEMATICA

Autore *JS. NEWTON*, *Trin. Coll. Cantab. Soc.* Matheseos
Professore *Lucasiano*, & Societatis Regalis Sodali.

IMPRIMATUR·
S. PEPYS, *Reg. Soc.* PRÆSES.
Julii 5. 1686.

LONDINI,

Jussu *Societatis Regiæ* ac Typis *Josephi Streater*. Prostat apud
plures Bibliopolas. *Anno* MDCLXXXVII.

the church that had granted an imprimatur and then prosecuted
Galileo for his 1632 work *Dialogue on the Two Chief Systems of
the World.*

The grantor of the imprimatur is perhaps as strange to us as
the permission itself. Samuel Pepys is best known these days as
a diarist, a bon vivant, and the man who established the fiscal
and supply side of the Royal Navy. Yet history does record that
he served as president of the Royal Society from 1684 to 1686.
Pepys was not a scientist but a man of universal interests, and
it was apparently as such that he was received into the Royal

Society, a purely scientific organization. He was a superb administrator and politically well connected. It thus served the purposes of the society to make him their leader.

A brief survey of Pepys's career gives a sense of the turbulent times in which Newton accomplished a work destined for the ages. In 1649, as a boy of 16, Pepys witnessed the execution of King Charles I by Oliver Cromwell's Parliamentarians. In 1660, as secretary to Admiral Edward Montagu, Pepys was on the boat that brought Charles II home from The Hague to reestablish the monarchy. One of this Charles's first acts was to issue the charter of the Royal Society in 1662. Pepys quickly rose to become Secretary of the Admiralty.

In 1679, Pepys was, on the basis of perjured testimony, accused of popery and treason. He was imprisoned in the Tower until he had gathered sufficient evidence to have the charges dropped. Strange to think of a man who was falsely accused of popery subsequently issuing imprimaturs. In 1684, the year in which he became president of the Royal Society, Pepys returned to his Admiralty job, where he remained until James II, the son of Charles I, was dethroned by William of Orange in 1688. At that point he gave up and retired. The age of Newton and Pepys witnessed the great plague of 1665 and the fire of 1666, which destroyed much of London. The century also was marked by England's many wars with the Scots, Dutch, French, and Spaniards. It was an unsettled time indeed.

The final section of the title page of the *Principia* gives us a bit more detailed information. The book was published in London under the auspices of the Royal Society. The name of that august body occurs no fewer than three times on the title page, an indication of its influence in the emerging age of science. The printer was Joseph Streater, and the book was on sale at numerous booksellers. The final entry, the year 1687, provides the basis for marking the 300th anniversary.

I now feel better educated about the *Principia*, and I have with this essay made my modest contribution to noting 300 years of Newtonian physics. As I write this, I am planning to toast Newton again tonight at dinner, but being in Kenya I shall do it by raising my bottle of Tusker Pilsner. Three cheers for Sir Isaac and the Royal Society, and thanks for keeps to Samuel Pepys. Hip, Hip, Hooray!

5

Albertus Magnus

A REPORTER WRITES of the happenings of the day, while an essayist waits a year to two or ten or a century or a millennium and then comments on an event. Thus, when the Einstein Memorial in Washington, D.C., was dedicated on April 22, 1979, I thought about writing a personal note on my feelings about the statue, but fortunately I waited, and some seven years later I am seated at the memorial on a sunny Sunday afternoon trying to get my thoughts together.

When the dedication took place, I had some negative feelings about the sculpture. First, it seemed clear that Einstein would have been unhappy with an over-three-times-life-size bronze figure of himself. He was so preeminently mental in outlook and modest in demeanor that it seems strange to honor him with 7,000 pounds of metal on a large stone base.

My second reservation was about the statue itself, which I had seen only in photographs. It is of a seated Einstein in baggy, crumpled clothes looking up from his pad with a tired, quizzical look. But there is something about Robert Berks's sculpture that in three dimensions has a totally different appearance and feel from the planar photographs. Wags may still refer to the piece as "the incredible hulk" or "Albert in a hot tub," but my plea to critics is to withhold judgment until they can interact—visually and perhaps tactually—with the original.

The third doubt I had centered on portraying the grandfatherly Einstein rather than the young man who managed, in spite of a full-time job in the Swiss patent office, to attack and solve most of the outstanding problems in the theoretical physics of his time. I felt that a statue of the young Einstein would serve as an inspiration to all those young people whose creativity is impeded and sometimes crushed by the dull, boring jobs they are forced to take to put bread on the table.

Well, that is what I would have written seven years ago. In the interim I have on a few occasions looked at the statue and marveled at how the warmth and humanity of Einstein literally glow from the stone and metal. I have rejoiced that our only national monument to a scientist depicts a kindly elderly gentlemen. Our capital has too many generals on horseback, and they are often ignored, but in this quiet grove of elm and holly trees on Constitution Avenue lies an inviting, comfortable place presided over by a friend in baggy pants.

The tourists do not come here in large numbers, as they do to the Lincoln Memorial. Rather, there is a steady stream of individuals and groups of two, three, and four. They come and gaze at the statue, or they sit on his lap or stand on his leg and address the great man. They, of course, photograph each other in those various poses. The statue of the great intellectual evokes so much of the common touch. The man's thoughts were cosmic, but his emotions were deeply human, and both aspects are apparent in the monument. A certain affection passes back and forth between the visitors and the statue.

The monument is on the grounds of the National Academy of Sciences, within sight of the Lincoln Memorial, where the Great Emancipator looks down in kindliness on the millions who throng to that center of national togetherness. And that nearness to Lincoln's statue reminds us of Einstein's avowal, "As long as I have any choice in the matter, I shall live only in a country where civil liberty, tolerance, and equality before the law prevail."

The Einstein Memorial also stands just a few hundred yards from the Vietnam Veterans Memorial, and that, too, reminds us of the great physicist's constant devotion to world peace.

The statue was financed with private contributions. The moving spirit behind the project was Philip Handler during his tenure as president of the National Academy of Sciences. Regarding the final funding, there is a persistent rumor whose source I have been unable to ascertain. The amusing but unconfirmed gossip asserts that the sale of the Audubon Folios by the National Academy was contemporaneous with the acquisition of the final monies for the Einstein Memorial. Individuals with knowledge of the matter are encouraged to write and enlighten me.

In any case, sitting by the statue is a pleasant experience, for it leads one to think of this strange man whose insights into the world of nature have done so much to influence the intellectual milieu in which we live.

I recently had the opportunity to read Albert Einstein's "first" paper in an article from the April 1971 *Science Today* by Jagdish Mehra. In 1894, Albert, then only 15, sent a letter to his maternal uncle Cäsar Koch. Included with the letter was a short paper, "Concerning Investigation of the State of Ether in Magnetic Fields." Hertz had recently demonstrated the generation of radio waves, which were believed to represent deformations of an immaterial ether that filled all of space. Einstein, in a short, nonmathematical work, talked about deformation of the ether and the velocity of propagation of electromagnetic waves.

Thus, the problems of physics that came to the fore in the special theory of relativity in 1905 were of concern to the 15 year old confronting the subject in 1894. At the time, Einstein was largely self-educated in physics and was unable to pass the examination to enter the Swiss Federal Institute of Technology in Zurich.

That is the way ideas seem to arise in the minds of truly gifted individuals. One is unable to account for them; they just appear. As Einstein himself noted, "Invention is not the result of logical thinking, even though the final result has to be formulated in a logical manner."

Well, the time has come to leave the statue. It still seems strange that a man whose thoughts were understood by so few should be such a popular figure to the steady stream of visitors. I think back to when I was nine years old and a book called *Einstein for Everybody* showed up at our house. It is inscribed to my father by the author, Herman Nurnberg. That is a great mystery to me, as my father had no familiarity whatsoever with physics. I read the book a few times while growing up. I suppose it influenced my emerging interest in physics and philosophy. That volume was my first contact with Albert Einstein, and I welcome the opportunity to pause here and think about his life a little more. There is definitely an upbeat feeling to this place, and I am grateful to all those who made it possible.

6

Fundamental Laws of Nature

SITTING IN AN AUDITORIUM at the California Institute of Technology one Friday morning in January, my thoughts began to drift. We were assembled to celebrate the 60th birthday of Nobel laureate Murray Gell-Mann, and the room was filled with people listening to talks on the future of research on the fundamental laws of nature. It was daunting to be in a room shared with so many brilliant thinkers in physics and mathematics, but there I was.

My thoughts, however, were floating back in time to past research on the fundamental laws of nature. I was thinking about a spring day some 44 years ago when Gell-Mann and I were in a large room on the second floor of the Sloane Physics Laboratory at Yale University. The course was "Physics 22a: Intermediate Laboratory Physics," presided over by Charles L. Clark. On the particular sunny afternoon I was remembering, several pairs of students were conducting an experiment described in the text as "Mechanical Equivalent of Heat by Puluj's Method."

The apparatus, set on a large, sturdy oak table, had a kind of traditional beauty, crafted as it was from wood and brass. The core of the piece consisted of two metallic cones, one inside the other. In a somewhat Rube Goldberg fashion, the inner cone was attached to a circular wooden disk, which was attached to a string, which in turn went over a pulley and was then connected to a large weight. The inner cone was filled with water, and a thermometer passed through the disk into the water. A handwheel was attached by a drive belt to both the outer cone and a mechanical counter, which registered the number of rotations.

The task of one member of the research team was to turn the handwheel fast enough to maintain a frictional force between the two cones just large enough to counter the gravitational force of the suspended weight, so that it neither rose nor fell. The second experimenter recorded the temperature and (with the aid of a stopwatch) the time. These two values enabled one to compute the heat produced. Given the counter reading, the magnitude of the suspended mass, and the diameter of the disk, one could also calculate the work performed.

This was a classical experiment indeed, a version of a series of measurements carried out by James Prescott Joule in the mid-1800s. Joule's results formed the experimental basis of the theory of the conservation of energy and the first law of thermodynamics. At Caltech, I was simultaneously hearing about future research on fundamental laws and thinking about past research on such laws.

In view of subsequent events, it was altogether appropriate that I turned the crank while my partner performed the more theoretical task of reading the temperature and time. Indeed, Gell-Mann's research on elementary particles forms an important part of the structure between the past and future of our understanding of the laws of nature, and I am honored to have collaborated on one of his lesser-known scientific endeavors. Moreover, I am proud to report that we did repeat Joule's earlier result to within experimental error.

Thinking back, I have asked myself whether I was aware in

early 1945 of the full potential of my young classmate. I must confess that at the time I was not, although by 1947 there was no doubt that my fellow student would somehow shake the foundations of theoretical physics. He had by then acquired what seemed to the rest of us an awesome depth of understanding of classical theoretical physics. As an undergraduate, he was already taking many graduate courses.

To return to Physics 22a, the text we used still occupies a place of prominence on my bookshelf. It is a second edition of *Laboratory Physics,* by Dayton Clarence Miller, issued in 1932, with an initial copyright date of 1903. Those experiments were based on a course that Miller taught at the Case School of Applied Science beginning in 1890. The last decade of the nineteenth century was the dividing period between the physics of Newton and Maxwell on the one hand and that of Planck, Einstein, and Bohr on the other. Our text began with mechanics and experiments on length, mass, time, acceleration, and gravity. Micrometers, planimeters, and balances were standard equipment in Physics 22a, and time was measured with tuning forks and metronomes. Digital display was unknown, and often one squinted to get the last significant figure. We were being treated to the full certainty of classical physics.

Regrettably, I am unable to recall many of the experiments, although there are memories of Young's modulus, Atwood's machines, and very sensitive galvanometers. The didactic methods of the day called for teaching physics by repeating some of the measurements that had defined the subject over the previous 200 years. As noted, the equipment itself had a certain classic beauty, being the work of artisans at lathes, milling machines, and lens-grinding wheels.

The next laboratory course we took brought us into the modern world. It was electronics, taught by Howard L. Schultz. Here, the aesthetics of the machinery was overshadowed by electron tubes, circuit boards, and the ever-present soldering iron. The microammeter had replaced the rotating mirror galvanometer. In the basement of the Sloane Physics Laboratory, the cyclotron was the harbinger of the future.

Indeed, the experiments that provided the basis of Gell-Mann's subsequent theoretical advances were performed on large particle accelerators. Much of the physics of the past 40 years has required the acceleration of subatomic particles to higher and higher energies. Such an approach mandates bigger and bigger machines, and devices of this kind continue to be central to our understanding of cosmology and the deep structure of the universe. In spite of this tendency, one could detect at the Gell-Mann symposium a search for theories that could be evaluated in less costly and more direct ways. To be sure, some present theories seem to lack any method of experimental check.

Given the great changes in perspective over the past century or so, it is remarkable how enduring the theory of the conservation of energy has been. When it apparently fails, as in beta decay, one simply invents a new particle, such as the neutrino, to take up the missing energy, and sure enough, the new particle is eventually demonstrated experimentally. That's the sort of thing that makes us biologists jealous of physicists, who find missing planets and missing particles by taking their laws seriously.

I am left with one residual question from all this rambling about the laws of nature: Who was Puluj? I have tried all my usual references on scientific biography and come up empty. The original text describing the experiment gives no references, and a conversation with an eminent historian of physics did not provide the answer. Unless my copy editors save me, I'm going to have to keep on wondering.

In the meantime, "Happy Birthday, Murray Gell-Mann."

II

7

Jesus, Moses, Aristotle, and Laboratory Animals

ISSUES SUCH AS the use of animals in research do not arise as isolated abstractions, as many "animal rights" advocates would have us believe, but are embedded in a cultural framework that goes back over 25 centuries. For contemporary American and European society, the moral structure of that framework has two principal foundations: the Judeo-Christian religious development and the rationalist-humanist school, which has its roots in Greek culture, particularly in the thinking of Aristotle and the Athenian Academy. While animal experimentation per se was not an issue in the classical world, contemporary attitudes on related issues established the basic approaches.

Central to Judaism is the Pentateuch, the first five books of the Bible, which are closely identified with Moses. The book of Leviticus contains elaborate descriptions of animal sacrifices: the

burnt offering, the peace offering, the sin offering, and congre-
gational sacrifices. The notion of animal sacrifices, while clearly
not the same as research on animals, nonetheless carries the
message that animals are killed for human benefit. Indeed, the
language of that practice has held to the present day: When
experimental animals are killed, scientific journals use the
phrase, "The animals were sacrificed."

By the time of the New Testament, animal sacrifice was no
longer predominant. Nevertheless, the concept of animals serving
as surrogates for ailing humans was still present. There occurs
a most dramatic story in the Gospel According to St. Mark:

> And they came over unto the other side of the sea, into the country
> of the Gadarenes. And when [Jesus] was come out of the ship, im-
> mediately there met him out of the tombs a man with an unclean
> spirit, . . .
>
> And always, night and day, he was in the mountains, and in the
> tombs, crying, and cutting himself with stones.
>
> But when he saw Jesus afar off, he ran and worshipped him,
>
> And cried with a loud voice, and said, What have I to do with thee,
> Jesus, thou Son of the most high God? I adjure thee by God, that
> thou torment me not.
>
> For he said unto him, Come out of the man, thou unclean spirit.
>
> And he asked him, What is thy name? And he answered, saying,
> My name is Legion: for we are many. . . .
>
> Now there was there nigh unto the mountains a great herd of swine
> feeding.
>
> And all the devils besought him, saying, Send us into the swine,
> that we may enter into them.
>
> And forthwith Jesus gave them leave. And the unclean spirits went
> out, and entered into the swine: and the herd ran violently down a
> steep place into the sea (they were about two thousand) and were
> choked in the sea.
>
> *Mark 5:1–2, 5–9, 11–13*

What is noteworthy is the willingness of Jesus to sacrifice 2,000
higher mammals to save one human being from his afflictions.
That attitude, which pervades both Judaism and Christianity,
comes originally from Genesis and the concept of man's dominion
over the animals:

And God said, Let us make man in our image, after our likeness: and let them have dominion over the fish of the sea, and over the fowl of the air, and over the cattle, and over all the earth....

And God blessed them, and God said unto them, Be fruitful, and multiply, and replenish the earth, and subdue it: and have dominion over the fish of the sea, and over the fowl of the air, and over every living thing that moveth upon the earth.

Gen. 1:26, 28

A central theme of biblical morality is the sacredness of each human life. Since man is the purpose of all creation, the notion of dominion follows in a natural way.

The tradition that has developed within the Roman Catholic branch of Christianity was explicitly expressed in *Moral Philosophy of Ethics and Natural Law* (1889) by Joseph Rickaby, S. J. The book was part of a series called Manuals of Catholic Philosophy:

Brute beasts, not having understanding and therefore not being persons, cannot have any rights.... The conclusion is clear. We have no duties to them—not of justice, as is shown: not of religion, unless we are to worship them, like the Egyptians of old; not of fidelity, for they are incapable of accepting a promise. The only question can be of charity. Have we duties of charity to the lower animals? Charity is an extension of the love of ourselves to beings like ourselves, in view of our common nature and our common destiny to happiness in God. It is not for the present treatise to prove, but to assume, that our nature is not common to brute beasts, but immeasurably above theirs, higher indeed above them than we are below the angels.... We have then no duties of charity, nor duties of any kind, to the lower animals, as neither to sticks and stones.... Much more in all that conduces to the sustenance of man may we give pain to brutes, as also in the pursuit of science.

Father Rickaby's point of view closely follows the writings of St. Thomas Aquinas. However, disagreement with this perspective on grounds of Christian charity and mercy was expressed by C. W. Hume in the *Dictionary of Christian Ethics* (1967).

A Protestant affirmation of Rickaby's view is found in Elmer Smick's entry in *Baker's Dictionary of Christian Ethics* (1973):

Not all forms of life, however, are equally sacred; there are levels of creation with man at the top. There should be no question of man's right to life above the animal world because he bears the Creator's image and has a spiritual nature.

The classical rationalist secular view lacks a single categorical statement, but its beginnings may be seen in the works of Aristotle, who as an experimental biologist had probably done hundreds of animal vivisections and dissections. He fully understood the importance of animal models in the study of human anatomy, physiology, and disease. In *Aristotle's Researches in Natural Science* (1912), Thomas East Lones concluded an account of Artistotle's investigations of animals with a short statement of Aristotle's views on man:

> Man is, for him, always a living being, an animal, but he is the highest representative of the whole series of living beings. He is distinguished from other animals by having a perception of good and evil, justice and injustice, and the like, and by his capability of reminiscence, this involving a process of syllogistic reasoning.

The extension of the Aristotelian perspective into the age of science is seen most forcefully in the writings of Claude Bernard. In his book *Experimental Medicine* (1865), he addressed the issue of animal use and gave what I believe is still the scientist's justification for work on higher mammals:

> Have we the right to make experiments on animals and vivisect them? As for me, I think we have this right, wholly and absolutely. No hesitation is possible; the science of life can be established only through experiment, and we can save living beings from death only after sacrificing others. Experiments must be made either on man or on animals. Now I think that physicians have already made too many dangerous experiments on man, before carefully studying them on animals. I do not admit that it is moral to try more or less dangerous or active remedies on patients in hospitals, without first experimenting with them on dogs; for I shall prove, further on, that results obtained on animals may all be conclusive for man when we know how to experiment properly. If it is immoral, then, to make an experiment on man when it is dangerous to him, even though the result may be useful to others, it is essentially moral to make experiments

on an animal even though painful and dangerous to him, if they may be useful to man.

Thus, our major religious and historical traditions contain strong elements not only permitting us, but in a certain sense urging us toward, the experimental use of animals for the benefit of man. Those traditions also direct us to be compassionate to the animals being used.

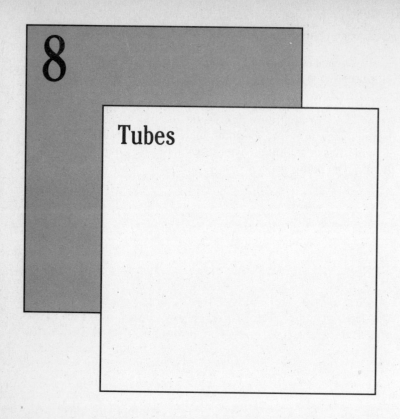

8

Tubes

WHILE DIGGING OUT a few references at the Yale History of Medicine Library, I chanced upon the name Gabriele Fallopio. Either the poetry of his eight-syllable designation or its association with anatomy was sufficiently intriguing that I jotted down on my pad "Check out Fallopio." A few weeks passed, and in reviewing the notes I came across the reference to the eminent Italian physician and pursued that information further. Fallopio was born in Modena in 1523 and died in Padua in 1562. During his short life, he made numerous contributions to our knowledge of human anatomy.

Fallopio entered research by a strange pathway. The *Dictionary of Scientific Biography* notes that he "began the practice of surgery but displayed so little aptitude for that subject—as demonstrated by the fatal outcome of a number of his cases—

that he soon thereafter abandoned it and returned wholly to the study of medicine."

The sixteenth century was without doubt the golden age of gross anatomy. The Renaissance was in full swing, and new attitudes toward the dissection of cadavers permitted an examination of issues and tissues that had lain dormant since Galen's work in the second century A.D. Andreas Vesalius' opus *De humani corporis fabrica* was published in 1543. The related volume *De re anatomica,* by Realdo Colombo, appeared in 1559. These anatomical researches centered in Italy, and Fallopio held the chair in anatomy at the University of Pisa, moving in 1551 to a similar position at Padua. He was clearly in the middle of where the action was.

Fallopio has, of course, been immortalized by the human oviducts becoming known as fallopian tubes. His name also lives on in the fallopian aqueduct, fallopian arch, fallopian canal, and fallopian ligament. Immortality, perhaps, came more easily in the 1500s. His career was marked by a number of other discoveries in human anatomy. He was truly a pioneer in describing the oviduct. The great anatomist used the word "tuba," medieval Latin for a trumpet, which describes the structure more accurately than the modern epithet "tube."

I might have passed over this newfound knowledge of Fallopio if I hadn't uncovered in these brief searches a discussion of a priority dispute between Fallopio and Bartolomeo Eustachi over the structure of the kidney. When I tried to enlighten my colleagues about this fascinating finding, they thought I was fabricating the story or that my good senses had gone down the tubes. But stay with me, dear reader—there really was a priority dispute between Gabriele Fallopio and Bartolomeo Eustachi.

The man for whom the eustachian tube was named was born in San Severino, Italy, in the early 1500s and died in 1574 while on the way from Rome to Fossombrone. From 1549 on, he served as Professor of Anatomy at the Sapienza. He is best remembered for his account of the auditory organ, tuba auditiva, which bears his name. Eustachi's studies on the ear appeared in his 1563 work *De auditus organis.* The anatomist is also remembered for the eustachian medulla and eustachian valve. His writings about the kidney appeared at about the same time as those on the ear.

Although Eustachi was an innovator and the discoverer of several new structures, he regarded himself as a follower of the renowned Galen, whose works contained several errors. The great early physician was quoted with dogmatic certainty for over 1,200 years. One of the subjects included in Galen's doctrine was anatomy, although that early collector of medical knowledge did not carry out human dissections. It is now clear from the published anatomical descriptions that he relied on data obtained from pigs and higher primates.

The reason that observations on humans were so sparse stemmed from an aversion to dissecting human cadavers, which seems to have dominated Christianity, Islam, and Judaism. A strong opposition to autopsy still persists in branches of these traditions. By reason of religious or civil law, dissections were prohibited for millennia.

With the Renaissance, the search for empirical knowledge seems to have overcome the taboos and proscriptions. Leonardo da Vinci (1452–1519) is considered by many to be the founder of physiological anatomy. His sketches are strikingly modern and could only have been the result of complete and detailed dissections, of which he claimed to have done 30. The drawings were not circulated at that time but were eventually published in 1784.

The best known figure in the development of anatomy was Andreas Vesalius (1514–64), whose magnum opus we have already noted. He served for five years as public prosector at Padua, which culminated in the printing of *De fabrica*. This epic work was a direct challenge to Galen's point of view on a number of points and broke new ground. After its publication, he tired of academic disputation at the tender age of 29 or 30 and left Padua to become court physician to Emperor Charles V. His chair in anatomy finally passed on to Gabriele Fallopio in 1551.

Bartolomeo Eustachi, several years younger than Vesalius, held a chair in Rome, and although he carried out dissections, he was unwilling to break with the doctrines of Galen. Thus, after Vesalius left the field of anatomy there developed contending schools of thought in Rome and Padua. The arguments between Fallopio and Eustachi had a deeper significance than just priority.

The flowering of gross anatomy occurred during the half century from 1530 to 1580 and was centered in middle and northern

Italy in a small number of universities. It is not clear why the religious prohibition of human dissection was eased at that time, but human anatomy, for both art and science, became central to the Renaissance. The sixteenth century in Italy is one of those times and places where knowledge leapt forward, opening new vistas for human thought. We are forever reminded of that epoch in the designation of the tubes, fallopian and eustachian.

9

Humans, Animals, and Physicians' Waiting Rooms

I WAS IN A SOMEWHAT adversarial situation with an "animal rights" advocate who was trying to convince me that if we gave up eating meat and conducting animal research, it would make us kinder, gentler people. To a biologist, something in these words didn't quite ring true. Having personally encountered the Cape buffalo and hippopotamus in Kenyan game preserves, I was well aware that these herbivores are among the most dangerous animals in East Africa. Their vegetarian diet does not induce gentleness. But a memory of an earlier reading placed the issue in a more human context, and I have looked up the appropriate references.

An article translated from a German publication called *Die Weisse Fahne*, volume 14, 1933, attributes the following quote to Dr. Goebbels: "Adolph Hitler does not drink alcohol or smoke and, moreover, is a vegetarian." The unsigned article goes on to say:

> Brother National Socialist, do you know
> ... that your Führer—out of persuasion and out of love for the world of animals—is a vegetarian...?
> ... that your Führer is the strongest opponent of any form of animal torment; especially of vivisection, which is the "scientific" torture of animals, and a heinous product of the jewish-materialistic school of medicine, about which he states: "In the National-Socialist state, this situation will end very soon"?
> ... that your Führer plans to prohibit all kinds of animal torture, especially vivisection, and by implementing this plan will save animals from their endless and nameless tortures and sufferings?

A subsequent article in the same journal notes:

> The Reich Press Agency of NSDAP announces:
> "The Prussian Prime Minister, Göring, issued a decree which prohibits vivisection of all animals in the entire Prussian territory beginning August 16, 1933. The Prime Minister has ordered the appropriate ministries to immediately draft a law which will impose high penalties on vivisection. Until this law comes into effect, all individuals who ignore this prohibition and arrange, perform, or otherwise participate in vivisection of any animals will be sent to concentration camps."

These quotes are presented here not to impose guilt by association, for none exists, but to note the absence of correlation between eating habits, views on vivisection, and ethical behavior. It is faint consolation to find that the most horrible imaginable crimes against humans in world history were committed by persons with a strong concern for animals.

Compassion for all living things is a noble enough sentiment, but in a rational context it must be structured by a number of factors, including our knowledge of evolutionary biology. We have to focus clearly on the fact that humans are different from the rest of the world of the living. This fact was unambiguously acknowledged in the religious foundations of Western culture: Animal sacrifice was mandated in the Book of Leviticus, and Jesus in the land of the Gerasenes readily sacrificed 2,000 swine to relieve the suffering of one tormented man.

From a secular perspective, modern biology reinforces the view that the emergence of *Homo sapiens* among the higher primates was a happening as radical as the very emergence of life itself. All life is to be cherished, but it is at the level of reflective thought and hominization that life becomes a categorical imperative. Not to recognize the profound differences between humans and all other life forms is to reject the evidence that meets our eyes at every minute. Failure to focus on the difference led Nazi Germany to enact antivivisection laws to protect animals and then to conduct cruel—often lethal—experiments on humans.

Cruelty to animals is totally unacceptable, but we do live in a natural world organized into food chains, in which herbivores eat plants, carnivores eat animals, and omnivores eat both. The consumption of animal food is not a human invention. If we use animals in this way, surely we are more than justified in using them to obtain the knowledge that will cure disease and alleviate the pain and suffering of humans.

Although concern for the suffering of all living beings is a virtue, I would urge those involved with the welfare of animals not to avert their eyes from the pain and suffering of humans. Once those problems have been addressed in full measure, we can focus more strongly on the lot of other creatures. Concern for the well-being of animals needs to be seen in the context of the human condition. In the Kenyan town of Lamu on the Indian Ocean, I once observed some British women running a shelter for infirm donkeys, yet many crippled and deformed lepers roamed the streets begging, with no one to care for them.

My interaction with animal activists dates back to my being appointed to serve on a National Research Council committee on the use of animals in biomedical and psychological research. Since my own research has been almost entirely confined to organisms from 0.5 to 100 μm in diameter, I was in the position of being a scientist without particular bias on the use of mammals and vertebrates. And indeed, I entered the study without bias. I had not formed an opinion on the issue of animal usage, and I have yet to firm up my views on how regulations should be keyed to taxa.

In the years of thinking about these issues I have become impressed that the so-called animal rights movement ("so-called" because the phrase "animal rights" is an oxymoron) is dominated by a one-issue ideology with an extremely narrow view. As such, it gains power because those of us broadly committed to a balanced morality are usually outorated and outmaneuvered by the ideologues. While the animals have highly vocal advocates, the mass of us who lead better lives because of animal research are silent. Our future health and welfare are being compromised by a small, highly active group of extremists and their supporters in Congress.

I have lately come to believe that the appropriate place for educating the public on this issue should be the physician's waiting room. Every such lounge should have available free pamphlets explaining how the diagnoses and therapies available to patients are the result of research protocols in which animal testing is an essential component. The use of animals in research has become highly politicized, and the action of legislative bodies depends on an educated public understanding the issues. Because physicians are the main link between the voting public and the outcome of biomedical research, it seems natural that their waiting rooms should become classrooms where the issues are explained. Perhaps the AMA and other medical, dental, veterinary, and pharmaceutical organizations could jointly prepare and publish the materials. Unless the public becomes better informed, we shall see our biomedical research programs in serious decline because of overrestrictive regulations that block innovative pursuits and make experiments too costly to conduct.

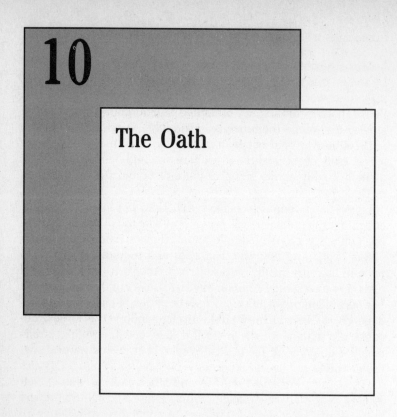

10

The Oath

THE HIPPOCRATIC OATH is one of those documents that is omnipresent in discussions of the foundations of medicine but seldom gets thought about in day-to-day medical practice. Therefore, when the subject came up during a classroom discussion on medical ethics, prompted by reading and analysis of the novel *Arrowsmith,* I was not able to quote it with accuracy. For the next class, copies were made available to everyone, and we began a critical reading.

The oath begins, "I swear by Apollo the physician, by Aesculapius, Hygeia, and Panacea, and I take to witness all the gods, all the goddesses." "Why," asked a student, "do all these Christian, Jewish, Moslem, and agnostic medical school graduates swear by all the pagan gods and goddesses of ancient Greece?" "Why, indeed?" I replied sagely, and we continued.

On reflection, I believe that answer might have been expanded. If we are to accept the ennobling features of taking an ancient oath, we must accord some sympathy to the culture of the authors. The crux of the oath is the Aesculapian's duty to the welfare of his patients, and that transcends most cultural differences.

The operative part of the oath begins:

> To consider dear to me as my parents him who taught me this art; to live in common with him and if necessary to share my goods with him; to look upon his children as my own brothers, to teach them this art if they so desire without fee or written promise; to impart to my sons and the sons of the master who taught me and the disciples who have enrolled themselves and have agreed to the rules of the profession, but to these alone, the precepts and the instruction.

I was taken with the veneration due one's teachers, but my students seemed less impressed than I with this part of the oath. What did impress them was the fraternal aspect of this profession. It was a calling, and the members were either born as sons of Aesculapians or became disciples. Some of this carries over today, in part because the training is so long and arduous and expensive and the ultimate responsibility so intense. In any case, the fraternal aspect persists, not mainly by oath but by shared experience. Indeed, there are aspects of residency that seem to have some kinship to fraternity hazing.

The next part of the oath gets into the substantive responsibility of being a physician:

> I will prescribe regimen for the good of my patients according to my ability and my judgment and never do harm to anyone.

This sentence is the core of the oath and is as valid today as it was in ancient Greece.

Further on in the vows, one comes to lines that are central to many disputes going on in contemporary American society:

> To please no one will I prescribe a deadly drug, nor give advice which may cause his death. Nor will I give a woman a pessary to procure abortion. But I will preserve the purity of my life and my art.

The sentence dealing with the deadly drug certainly rules out the physician's being part of murder by poisoning, but it would also seem to unambiguously rule out euthanasia. It is thus relevant to our attitudes on the care of the terminally ill.

One wonders if an Aesculapian provided the hemlock for Socrates. The savant's last words as reported in Plato's *Phaedo* were: "We owe a cock to Aesculapius; pay it without fail." Commentators have interpreted these lines as the poor man's thank offering to the god of healing, for Socrates was being cured of the pain of living. Still and all, one wonders if these words might be a final satirical criticism of those who violated their oath in providing the state with poison to carry out the official murders. One also thinks of physicians who give lethal injections in those states where capital punishment is so mandated. Somehow, it would be well to separate the offices of physician and Lord High Executioner.

The word "pessary" comes from the Greek πεσσόσ, whose original meaning was an oval pebble. It came to refer to any device worn in the vagina. This provides some evidence of how some abortions were carried out, but in any case the oath seems definitely antiabortion. Once again, we are faced with the issue of using an ancient document that emerged from a culture quite different from our own.

The next affirmation precludes the physician from doing surgery:

> I will not cut for stone, even for patients in whom the disease is manifest. I will leave this operation to be performed by practitioners (specialists in this art).

This would appear on the surface to be simple trade unionism. Indeed, it even suggests referral. That the physician could diagnose the disease was clear, but specialization had begun.

The physician then declares:

> In every house where I come I will enter only for the good of my patients, keeping myself far from all intentional ill-doing and all seduction, and especially from the pleasures of love with women or with men, be they free or slaves.

Clearly, in the time of Hippocrates most calls were house calls, and good faith demanded that the physician not take advantage either of patients in a weakened and confused condition or of their households. It still does.

The final directive part of the oath deals with doctor-patient confidentiality:

> All that may come to my knowledge in the exercise of my profession or outside of my profession or in daily commerce with men, which ought not to be spread abroad, I will keep secret and will never reveal.

The phrasing does allow considerable discretion on the part of the physician, but nevertheless, it speaks to the issue of privileged information and is designed to protect the rights of the patients. The oath ends with the rewards and penalties of compliance with or violation of the oath.

The classroom discussion dealt in part with the issue of whether we ought to use such an ancient document, parts of which are not endorsed by present-day swearers. If oath taking is to be a serious act, it is unreasonable to follow it with a series of exceptions. Is continuity with the ancient world so important that one should swear falsely, particularly if doing so involves violating another ancient document?

It is clearly discomforting to reexamine accepted practices, and an operational ethic limits the effort we can accord to such reexamination. Nevertheless, morality also appears to have temporal components, since it is geared to technology and changing knowledge. I suspect that it would be healthy for the medical profession to devote some time to discussing anew the Hippocratic Oath and asking whether it should be retained.

11

Autism and Authority

IN THE UNITED STATES TODAY, major intellectual and scientific decisions are often made by judges and jurors who are not free to step back and endlessly debate conflicting viewpoints but must render decisions sharply and definitively, often causing appreciable amounts of money to change hands. Such a case was that of Daniel Kunin, plaintiff, versus Benefit Trust Life Insurance Company, defendant, an action brought in United States District Court, Central District of California, Judge Irving Hill presiding. Opinion was rendered on September 19, 1988.

The crux of the decision, which will likely be of major importance in medical-legal annals, is embodied in the first sentence:

In this opinion the Court, in an apparent case of first impression [i.e., no precedents], decides that autism is not a "mental illness" within the meaning of an exclusionary clause in a group health and medical insurance policy.

The plaintiff was covered by a policy issued to his employer by the defendant. In 1986 the plaintiff took his son to the UCLA Neuropsychiatric Institute. There the child was hospitalized for 30 days and diagnosed as having "organic brain dysfunction... and the syndrome of autism secondary to the first diagnosis."

The bills for hospitalization and treatment came to $54,696.96, and the plaintiff submitted a claim for that amount. The insurer argued that the policy limits benefits for "mental illness or nervous disorders" to a maximum of $10,000 in any calendar year. The plaintiff sued for the unpaid portion of the claim— $44,696.96.

The judge in his opinion described the nature of the proceedings:

> Much of the trial concerned itself with expert testimony as to the definition and description of autism and its etiology. That testimony established the facts set forth below.
>
> The syndrome of autism falls within the category of pervasive developmental disorders. The syndrome is defined by its symptoms. The essential features are a lack of responsiveness to other people, gross impairment in communicative skills, and bizarre responses to various aspects of the environment.... These symptoms usually appear within the first 30 months of life.

This section of the proceedings then concludes:

> Research into the causes of autism has also led to the conclusion that the syndrome is not environmentally or psychologically based. There is a consensus among experts that the syndrome of autism is not caused by environmental trauma or childhood relationships with parents or others. It is clear that autism cannot be treated by traditional psychotherapy.

Judge Hill thus resolved, in a legal sense at least, a dispute with a history of more than 40 years. In 1943, Leo Kanner published an article describing 11 children who in his view came "into the world with an innate inability to form the usual biologically provided affective contact with people." He thus defined a syndrome, early infantile autism, which was the title of his 1944 paper in the *Journal of Pediatrics*.

A controversy quickly arose as to whether infantile autism was due to inadequate mothering or an inborn deficit. The inadequate mother hypothesis, a psychoanalytic etiology, achieved strong support from the famous child psychologist Bruno Bettelheim. Bettelheim's views are summarized by Austin M. Des Lauriers and Carole F. Carlson in their 1969 book *Your Child Is Asleep: Early Infantile Autism:*

> Bettelheim himself, who never truly clarifies whether he is talking about primary autism (early infantile) or secondary autism (childhood schizophrenia), chooses to view the parental environment as responsible for the fact that "vitally needed experiences do not occur" in the early life of the autistic child. Basing his view on the many years of clinical experience he has accumulated in working with a variety of disturbed children, Bettelheim reached the conclusion that the autistic condition in a child is directly consequent to the wish of the mother that this child "did not exist." The child, very early, according to Bettelheim, senses somehow this basic rejection by the mother, and tries, in defense, "to blot out what is too destructive an experience" for him. In so doing, the child, his back turned on the world, so to say, remains unavailable to it. In protecting himself from the destructive designs of his mother, he ends up by defending the deprived and "empty fortress" of his life.

Bettelheim's view was not an abstraction. I have interviewed a woman who some 25 years ago traveled all over the country seeking help for an autistic son. She ended up in Dr. Bettelheim's office being told that the problem was that she didn't love her child. She responded by opening up a school for autistic children and became a national authority on institutions that deal with autistic children and adults.

Given Judge Hill's clear and unequivocal statement, one wonders how Bettelheim's authoritarian foolishness could have held a central position for so long. Extremist Freudian thought had apparently become so much the paradigm among some professionals that any assault on despairing parents was considered legitimate.

The court turned to another question that also makes this opinion precedent-setting:

Turning to the question at bar, one preliminary question must be addressed. The words in issue in this case, "mental illness," can conceivably be characterized as a scientific or technical term. The Court must therefore decide whether the words should be given an exclusively scientific or technical definition as opposed to a "lay" definition.

The Court has determined to reject a scientific or technical definition. One cannot ignore the fact that the words in question were used in an insurance policy which was written by nonscientists, approved by nonscientist insurance commissioners, and purchased by lay persons for the protection of lay persons.

It has been long-established that insurance policy language and terms should be construed in accordance with the plain and ordinary meaning that a lay person would ordinarily attach to them rather than in their technical sense. In this Court's view, the policy term in question here should also be given the plain and ordinary meaning that a lay person would ordinarily attach to it.

Within that context the court drew its conclusion:

As the evidence indicates, mental illness is often thought of by lay persons as having nonphysical, psychological causes, in the Freudian sense, as opposed to an organic basis. Where dysfunctions of the brain derive from an identifiable organic basis, as in the case of brain cancer or Alzheimer's disease, the condition would not commonly be understood as mental illness. A related factor which can distinguish mental illnesses is their causation by environmental factors, such as traumatic experiences or childhood relationships. Autism exists throughout the world. It is not caused by environmental factors and it is unaffected by manipulation of environments. Its incidence and characteristics remain constant across socio-cultural environments. . . .

To summarize, the Court concludes that defendant's inclusion of autism within the limitation clause covering "mental illness" was not a reasonable interpretation of the contract and the plan. As a result, defendant has arbitrarily and capriciously denied plaintiff's claim for benefits in violation of 29 U.S.C. §1132, and is liable to plaintiff for those unpaid benefits plus prejudgment interest.

This case is now in appeal. As it stands, the precedent will be of enormous importance not only in cases of autism but also in cases of numerous other diseases where psychoanalytic hubris, often many years out of date, is depriving many insured people of funds for the care of ill relatives. As Judge Hill's last footnote makes so clear, we still have an enormous challenge ahead of us:

> There is no cure for autism. At present, the treatment consists of a combination of medication, to control the physical problem to a limited extent, and special education to help the patients make use of their strengths and function around their weaknesses.

12

The Missing Volumes

I AM AN UNREPENTANT DEVOTEE of libraries. I have never out-grown habits acquired as a youngster, wandering down to Adriance Memorial Library in Poughkeepsie, N.Y., on Saturday afternoons and spending endless hours reading about whatever came to mind. It was truly a window on the world. Library books have a characteristic smell, a feel—one might almost say an aura. The edges of pages crack in characteristic ways. There is something sensual as well as intellectual in being surrounded by aged books. I also love the silence of the stacks, for aisle-wandering is a lonely occupation.

On rare Saturday afternoons, I meander down the street to the Fairfax Library for the simple pleasure of watching young people interact with the printed word. They are a diverse group, and it is good to know that the art of reading is not lost among them.

And there is always some new subject to explore, some unread work to look into. Now, to enrich one's intellectual life even further, on-line information and bibliographic data bases have come to enhance the existential pleasure of time spent as a bibliophile.

Thus, on weekends, I now occasionally find myself in the depths of the library at the Clinical Center in Bethesda, Md. This is not the great archival National Library of Medicine but is a well-run, open-shelf working library for scientists maintained by the National Research Resource Center of the National Institutes of Health. Although the book collection is not huge, the journal collection is very extensive, and I usually find what I'm looking for. And as a bonus, the data base GRATEFUL MED (which accesses the NLM's collection of information) is available on-line.

The library is not heavily used on weekends, so quiet and solitude abound. Sometimes in the large downstairs stacks there will be only three or four people. One of the very best features of such a repository is that one never knows what line of thought will result from purposeful or chance encounters with the surrounding materials. Such are the rewards of browsing.

So it was one recent Saturday morning when I sought an article on brain birth and personal identity in the 15th volume of the *Journal of Medical Ethics*. This paper, published in 1989, dealt with some material that has been much on my mind lately. The volumes of this journal on the shelf ceased at 1985, so a stop at the computer terminal was indicated. After I typed in the name of the journal, the message on the screen indicated the volumes that were present and the message, "Subscription for 1986 to 1990 suspended by the Library. Latest received: September 1991."

It is unusual for a research library to allow a lacuna in the journal shelf collection. Therefore, the cryptic message about the *Journal of Medical Ethics* was even more intriguing and brought to mind all sorts of conjectures, ranging from the frivolous to the conspiratorial. For, by almost anyone's judgment, the years 1986 to 1990 were the period of greatest concern over ethics in the brief history of the National Institutes of Health.

Toward the beginning of the period, in April 1986, an article appeared in *Cell,* with David Weaver, Thereza Imanshi-Kari, and

David Baltimore among the authors. The era ended with the National Academy of Science appointing a committee that finally made a recommendation to have required courses in ethics and conduct within the regular curriculum for science students. In the interim, many senior American scientists and some members of the upper levels of administration of the NIH did not always respond to the ethical injunctions of putting truth above persons and integrity above the vested interests of the multibillion dollar programs of the scientific establishment. An appreciable number of scientists, with little knowledge of the case under discussion, went public to support David Baltimore, even though this involved quashing the career of Margot O'Toole, the whistle-blower. It was this blatant case of protecting one's own that shattered the long-standing affection of the public for the scientific community. The *New York Times* wondered if it were a scientific Watergate.

The period covered by the missing volumes also overlaps the lengthy, messy, and seemingly unending investigation of the strange goings-on at the laboratories headed by AIDS researcher Robert Gallo. All of this involves supposedly purloined viruses, interests in patents of enormous financial value, and a run for the Nobel. An NIH draft report proposing that Gallo's actions "warrant significant censure" is still being withheld because of objections posed by the director of the NIH.

A number of other cases around the country are suggesting one or another level of scientific misconduct in research funded by and presumably monitored by the NIH. Several of these related to events of the years 1986 to 1990. It has been a difficult period for an organization that has made monumental contributions to scientific knowledge and medical practice.

Like many scientists, I have agonized over the Baltimore and Gallo affairs. I must plead guilty to being naive enough to think of science as a quest for truth and understanding rather than a race for fame, money, honors, and adulation. Providing that one can make a reasonable living, the job of discovery is reward enough for a life spent in the most enjoyable of pursuits. That may sound unsophisticated in the extreme, but it happens to be the case.

But now I have an explanation for all that has gone wrong.

Five years of making decisions without the availability of the *Journal of Medical Ethics* was just too much. It left power to the pragmatic and the political. Under such circumstances, the system fell into moral disarray.

One is left wondering why the subscription was suspended (not the passive "lapsed," but the active "suspended"). Was it a budgetary decision or a policy decision? Why, for example, did the NIH retain *Helvetia Chimica Acta* and drop the *Journal of Medical Ethics*? Is there a more conspiratorial reason? Was this regarded as a dangerous publication? Did it fail the litmus test on abortion? Who knows?

But rejoice, friends, the subscription has been reopened, and knowledge of ethics again freely flows to Building 10. I have two modest suggestions for the powers-that-be at the NIH:

1. Obtain copies of the missing five years' volumes.

2. After these are installed in the stacks, review all decisions made since 1986 that have an ethical component.

Back in the stacks, I get to track down my next reference. It is a classic article on chromatography in *Helvetia Chimica Acta*. The world is very complex.

13

The Smoking Gun

NOW THAT SCIENTIFIC INTEGRITY has become a major public issue, I am tempted into going to the closet and dragging out a skeleton that has been gathering dust for many years. I would have done this earlier, but the current climate of opinion seems more favorable.

In 1974 I attended an international ecology meeting in the Netherlands. I was discussing some scientific theories with a young Belgian scientist, who pointed out that I would appreciate the ideas in a new book, *L'Anti-hasard* by Ernest Schoffeniels, which had just appeared in print. He offered to bring his copy to the next session to show me. When he appeared with the volume, he thought I would be particularly interested in Chapter 4.

I must confess that my knowledge of French, stemming from

two years of high school instruction, is, to say the least, weak. Even so, I was surprised at the ease with which I read the material. My surprise turned to amazement when I realized that many of the words were a Gallic version of sections of a book I had written, published in 1968.

To let the reader judge for him- or herself, I shall list sections from Chapter III of *Energy Flow in Biology* and the corresponding sections of Chapter 4 of *L'Anti-hasard*. *Energy Flow* (EF) lists 13 biological generalizations. *L'Anti-hasard* (LA) lists 14 propositions that are "Les bases d'une biologie théorique."

LA 2ᵉ Proposition: Dans tous les systèmes biologiques le composant principal est l'eau.
EF I. The major component of all functioning biological systems is water.

LA 3ᵉ Proposition: Dans tous les systèmes biologiques, les atomes intervenant dans les liaisons convalents sont: C, H, N, O, P, et S.
EF II. The major atomic components in the covalently bonded portions of all functioning biological systems are carbon, hydrogen, nitrogen, oxygen, phosphorus, and sulfur.

LA 5ᵉ Proposition: Dans tous les systèmes vivants, le poids sec est essentiellement déterminé par la présence de protéines, de lipides, de glucides, et d'acides nucléiques.
EF III. Most of the dry mass of functioning biological systems consists of proteins, lipids, carbohydrates, and nucleic acids.

LA 8ᵉ Proposition: L'information biologique est structurale.
EF V. Biological information is structural.

LA 9ᵉ Proposition: Dans tous les systèmes biologiques, le flux d'énergie est accompagné de la formation et de l'hydrolyse de liaisons phosphate.
EF VI. The flow of energy in the biosphere is accompanied by the formation and hydrolysis of phosphate bonds, usually those of adenosine triphosphate.

LA 12ᵉ Proposition: Dans les conditions actuelles, le maintien de la vie est une propriété d'un écosystème plutôt que celle d'un seul organisme ou d'une seule espèce.

EF VII. Sustained life under present-day conditions is a property of an ecological system rather than a single organism.

LA 11ᵉ Proposition: Il y a un type universel de structure membranaire utilisé dans tous les systèmes biologiques.

EF IX. There is a universal type of membrane structure utilized in all biological systems.

LA 13ᵉ Proposition: Toutes les populations de systèmes biologiques qui se reproduisent donnent naissance à des mutants phénotypiques, reflétant des altérations du génotype qui statistiquement sont irréversibles.

EF X. All populations of replicating biological systems give rise to mutant phenotypes that reflect altered genotypes.

Well, I could go on and on, but by now the reader is no doubt keenly aware of how my knowledge of French took a quantum leap forward when I encountered this book. I then searched the footnotes for a reference to my work. None was to be found. I returned home and contacted the publisher of *Energy Flow in Biology,* Academic Press.

My publisher informed me that French copyright law was notoriously weak (the book was published by Gauthier-Villars Éditeur). The only consolation they could offer was to tell me that the author of *L'Anti-hasard* was negotiating with them for an English edition; they would immediately withdraw from these negotiations. They also wrote to inform Professor Schoffeniels of the problem. He denied any plagiarism.

At that I let the matter stand. Ernest Schoffeniels, according to the title page, was Professeur de Biochimie à l'Université de Liège and Professeur associé à la Duke University, Caroline du Sud (I had always thought that Duke was in Caroline du Nord). To the best of my knowledge, he was a respected biochemist. After thinking about it for 15 years, I have decided I was in error in dropping the matter.

The present mess we are witnessing in science over postdated data, allegedly purloined viruses, created data, and the like indicates that we cannot give up our hair-shirt consciences if science is to survive. It is not only our option to blow whistles when we learn of wrongdoing, it is our duty. The entire structure of science is a social enterprise in which we stake our careers on the work of others. Anything less than an absolute policy of whistle-blowing endangers the enterprise. I have waited 15 years to blow this whistle, mea culpa, and (to mix metaphors) I have just shown you some of the smoke from this smoking gun. It is quite an unpleasant thing to do. If I were a beginning scientist, it would be a daring thing to do.

We have a strong social aversion to whistle-blowers. More and more I feel that this is a serious mistake. We need an honor code in science. We need to be protected from an establishment that takes care of its own. We have to stop compromising with an *omertà* (conspiracy of silence) attitude.

Well, my whistle is blown, and I feel better. Have I whetted any other whistles?

III

14

Just Say No

THE SUDDEN DEATH of Baseball Commissioner A. Bartlett Giamatti within a week of President Bush's announcement of the war on drugs brings back some personal memories. It also evokes some thoughts on substance abuse and the magnitude of the problem of dealing with our addictive habits. A series of events during the years 1981 to 1986 occasioned several conversations between Bart and me dealing with alcohol control, tobacco use, and related matters. There were other things both of us would have preferred to discuss, but our job responsibilities had brought us to these matters.

Bart was, of course, the president of Yale University and as such was seriously concerned with all aspects of campus life. I was master of Yale's Pierson College and charged with the well-being of some 400 young people who were experiencing new freedoms amid the full transition to adulthood.

During 1982, '83, and '84, the legal drinking age advanced each year—from 18 to 19 to 20 to 21. The successive restrictions were imposed on the states by the allocation of federal highway funds and were a response to the alcohol-based carnage on the national roads. I think that few people outside the residential academic community are aware of the profound changes in campus life that resulted from the shifting drinking age. There was a large movement of social functions from campuses to fraternities, sororities, and rented party facilities.

Giamatti and I took quite different approaches to these problems. He opted for a strong enforcement policy to eliminate consumption of alcohol by underage students at campus events. I worried that a draconian attitude on the part of university officials would force alcohol-consuming young people back onto the highways and into partying situations unattended by adults with *in loco parentis* authority.

The president's office at Yale, where we met, is a small, modest room on the second floor of Woodbridge Hall. The rectangular cubicle has a single large window looking out on Beineke Plaza. I stress the size of the office because the limited volume of air impressed on the visitor the full measure of the president's smoking habit. In addition, the smoke was usually increased by the presence of the assistant to the president, a man who shared his addiction. On occasion, I felt that the clarity of my arguments was obscured by the dense haze that filled the room.

At one point when our differences were becoming uncomfortably great, I opted to change the subject by interjecting, "Bart, I worry that you smoke too much." Giamatti, who was one of the quickest-witted persons I have ever met, shot back, "Only when you're in the office." Alas, his verbal rejoinder was more clever than accurate; he smoked too much all the time.

Now, in general I am not given to hortatory approaches. In dealing with students, I try to resist being judgmental. But on the smoking issue I cannot resist helping young people break the habit. As a former smoker, I had studied the Hammond report of the American Cancer Society with great interest. This document on the morbidity and mortality associated with smoking is, to my mind, one of the great documents of modern medicine. I plan to write a tribute someday to its author, E. Cuyler Hammond. I suppose that over the years I have helped 10 or more students give up smoking, and I count that effort one of the more significant things I have done in my academic life.

It is one thing, however, to expound views about smoking to an undergraduate and quite another to undertake the education of the distinguished president of a major university. I opted for an intermediate pedagogic position, made a photocopy of my well-worn pages of the Hammond report, and sent it along with a brief note "for your interest." A note of thanks was returned without further comment.

During the following years and through a number of interactions, I was constantly surprised that a person of such force, personality, brilliance, and obvious accomplishments couldn't just say no. Giamatti had a sense of rectitude, even a puritanical air about doing the correct thing. When he became commissioner, I felt that baseball was in for a very strict set of regulations and enforcement in the area of drug use, and in his short term as chief baseball administrator, we saw the kind of toughness we all had known would be exerted in rule enforcement. I have a sense that he must have felt great remorse about a continuing addiction that violated his personal ethos.

In his final months in office, C. Everett Koop, surgeon general of the United States, made an eloquent case for the addictive nature of tobacco and the national health menace of smoking. His statistics and conclusions were impressive. Smoking is a powerful addiction and a major public health problem around the world. The anecdotal information, however, is even more moving than the statistical. Particularly moving is the sight of powerful men and women whose wills have been compromised by this drug.

A Dutch colleague who spent time in a slave labor camp in World War II told me that most of the deaths in that factory camp were of smokers who would trade food for cigarettes, both of which were distributed in meager amounts. He told of men who had been professionals and leaders of their communities fighting with each other for the cigarette butts their guards threw away.

I recently attended a national committee meeting where the

distinguished chairman was unable to be a full participant in the discussions because we met in a building where smoking was not permitted. He was constantly sneaking off to get his fix. It was a telling performance.

We have no way of knowing whether heavy smoking was a major cause of Giamatti's coronary seizure. Hammond's data certainly show that smoking increases the probability of such episodes. We do know that smoking a pack a day or more doubles the mortality for most causes of death. We also know that there are millions of people who would like to stop smoking and seem unable to do so.

These observations on the legal drug tobacco have profound implications for the so-called war on drugs. They remind us that to discourage beginning drug use, it is important to create a society in which reward comes from the normative aspects of life. The cure for addiction seems to be never to become addicted. That aspect requires not a police force but a value system in which other things in life are of more value than instant gratification.

The last time I saw Bart Giamatti was in the Grill Room of the New York Yale Club. It was a casual meeting, and we chatted briefly. He had just been made baseball commissioner. He said he had heard I also had changed jobs and wondered about it. I lightheartedly replied, "They made me an offer I couldn't refuse." With a twinkle in his eye, his rejoinder was, "Me too!"

15

The Fromagification of America

I WAS SITTING in a favorite Fairfax, Va., bistro perusing the menu when I noticed that of the eight varieties of sandwiches and seven varieties of gourmet burgers offered, only four—a mere 27%—were prepared and served without a cheese covering or cheese melt. Deployed by the chef were Wisconsin cheddar, mozzarella, Monterey Jack, and Swiss. Even the all-American steak sandwich was "topped with grilled onions and melted mozzarella cheese."

I could have ignored this abundance of cheese as a particular passion for solidified milk products on the part of the management of this establishment, but I have been collecting observations. A year ago on temporary assignment in Palo Alto, Calif., I first began to notice the difficulty of ordering food pressed between two pieces of bread without accompanying cheese. Even the egg on a bagel at Burger King came with melted cheese. A phenomenon

67

common to such diverse locales as Fairfax and Palo Alto must
have some deeper meaning. Since then, I have added observations in New Haven, Conn.; Santa Fe, N.Mex.; and Racine, Wis.,
and I am now convinced that I have stumbled on a genuine
discovery in culinary sociology. The final confirmation came
when riding an AMTRAK train from Washington, D.C., to New
York. Five of the eight sandwiches sold in the café car contained
cheese in various stages of the solid-liquid transition.

First, a word about great sociological and historical changes.
They often creep up on us slowly, and we are unaware of what
is happening until one day we wake up and, *voilà*, the Roman Empire has declined and fallen. The residents of Neapolis
(present-day Napoli) in the mid-400s A.D. were almost certainly
not thinking about decline and fall. They were busy perfecting
mozzarella. Thus, it is especially challenging to find a great social
movement in progress so that it can be carefully studied in situ.
I would like to date the beginnings of fromagification (the great
increase in the amount of cheese in the average diet) to around
1955. I suspect but cannot prove that it may have something to
do with the termination of the various state laws against the
coloring of oleo-margarine and the health concerns over fats in
the diet, all of which left the dairy industry with gigagallons of
high-butterfat milk.

In 1955, McDonald's was becoming a national institution, and
the hamburger—not the cheeseburger—was the lead sales item.
By 1980, Quarter Pounders with cheese were available at any
time, but a Quarter Pounder without cheese was a special order.
Back in 1955, many establishments offered the Reuben, a combination of heated corned beef, sauerkraut, and Swiss cheese,
but few other lactiferous sandwiches were available, with the
exception of the ubiquitous ham and cheese. Now, as has been
indicated, it is sometimes difficult to find a cheese-free sandwich
on the menu.

What is of particular interest about this change in cuisine is
that it has occurred at the same time as the great surge of interest
in health foods, low-fat diets, and low-sodium diets. With this in
mind, let us examine the compositions of some cheeses, based
on a 50-gm quantity—the amount that probably goes on a sandwich (see table). The fats are approximately 60% saturated fatty

Cheese Compositions (50-gm Portions)

	Protein (gm)	Fat (gm)	Sodium (mg)	Cholesterol (mg)
Cheddar	12.5	16.1	350	50
Swiss	13.8	14.0	355	50
American	11.6	15.0	568	50

acids, and a 50-gm portion of cheese adds about 200 calories. In a diet-conscious country, I would not have imagined that cheese would be a food of choice, yet it has become a food of fashion as well as a fast-food staple. We gratuitously add to many sandwiches about 350 mg of sodium (the equivalent of about 900 mg of salt), about 10 gm of saturated fatty acids, 50 mg of cholesterol, and 200 calories. In addition, a substantial number of people with lactose intolerance have to watch their sandwiches with particular care when dining out.

One is left with the question of why a society has radically changed its eating habits in a systematic way that counters the general wisdom about health and nutrition. One can only speculate, and such speculations can be exceedingly uncertain. I recall a statement rumored to have been made by the late General Charles de Gaulle: "Two hundred forty-eight cheeses—who can govern a country with two hundred forty-eight different cheeses?"

I can point to some parallels to the sandwich phenomenon. The Americanization of pizza, which began in 1935, introduced people to the idea of food with melted cheese. Indeed, a pizza can be viewed as an open-faced tomato melt sandwich. The success of pizza cleared the way for fromagification.

The rise of European cuisine in post-World War II America led to familiarity with quiche, fondue, and other cooked cheese dishes. The northward immigration of Mexicans into the United States brought another cuisine rich in melted cheese. At least one technological advance contributed to the utilization of cheese: The microwave oven created a new opportunity to put the cheese in its most solid form in a sandwich and then melt it.

These partial explanations hardly have a ring of authority about them, but styles in foods must clearly involve economic and cultural factors and only mildly reflect nutritional factors. I have observed associates at lunch forgo, for health reasons, usual meat dishes and sit down to a chef's salad that included

- Hard-boiled eggs—a major source of cholesterol
- Salami—a collage of salt and saturated animal fat
- Ham—a rich source of salt
- Olives—a tasty combination of salt and calories
- Mayonnaise—a source of caloric energy gallon-for-gallon comparable to diesel oil
- Cheese—a food whose virtues I have already discussed

Now, I do not wish to oppose chef's salad. On occasion and in moderation I eat it myself. But I do so because I like it, not because of any supposed nutritional virtue.

To return to fromagification, I must recall my recent five years of eating in a college dining hall, where each meal had a meat, fish, and vegetarian choice. The vegetarian alternative was almost always smothered by melted cheese in some form or other. On reflection, this was nutritionally sound. The herbivores among us can use a source of animal fat and protein. Cheese eating enables them to consume such foods without having to consider such disturbing thoughts as the dairy industry's being the source of veal.

Back to the menu I was studying: The special of the day was Cajun Chicken Sandwich, described as breast of chicken sautéed in Cajun spices and served on a bun. I ordered it, and it arrived with melted cheese, which hardly seemed like a Cajun spice. So, I conclude, the presumption is cheese unless otherwise specified.

16

Killer Cheese

NUTRITION REMAINS ONE of the most enigmatic subjects in modern health care. It is doubly perplexing: First, because reliable information on human diet and health is so difficult to obtain by experiment or epidemiologic analysis, and second, because of the public's unpredictable responses to the available information. A modest case study illustrates these features.

Among the dietary caveats that have emerged over the past 20 years, three have stood out as having achieved consensus support among nutritionists. These are

1. Moderate the amount of fatty acid in the diet, with particular attention to minimizing saturated fatty acid. This recommendation comes with particular force from students of heart disease and cancer.

2. Moderate the amount of dietary cholesterol. This advice may be appropriate only to certain genetic subgroups, and the relationship between dietary and blood cholesterol has been questioned. Nevertheless, the recommendation has been advocated persistently for many years by those involved in preventing and treating arteriosclerosis.

3. Moderate the amount of salt—and more specifically, the amount of sodium—in the diet. Again, this may apply only to a subgroup with familial hypertension but can also apply to a much broader group.

These three warnings have received widespread notice and are conversational items at teas, cocktail parties, and in locker-room discussions. They are pervasive influences on the modern American mind at a time when we have become very health conscious and keenly aware of possible influences of diet on health.

Given this emphasis on health in eating, it is very strange that we have entered into a culinary mode that regularly mandates placing on one's food slices of material very high in saturated fatty acids, cholesterol, and salt. For if the perceived wisdom about nutrition is correct, these slices are truly "killer cheese" when we apply dietary data to mortality statistics.

This perception about cheese slowly works its way into the public consciousness in strange ways. In an article in the *Hartford Courant* (August 15, 1991), there is a discussion of Yale–New Haven Hospital's divesting itself of the tobacco stocks in its portfolio. The article reports on a meeting of the general counsel of Phillip Morris with the appropriate hospital committee. According to the *Courant,* the lawyer "went on to compare the cancer risk from smoking to that of eating cheese for high cholesterol adults." Given what we know about smoking, this is killer cheese indeed!

We are left with the question of why a sophisticated eating public turns its back on cream in opting for 1% and 2% milk, then turns around and puts that same fat and cholesterol, augmented with salt, on a vast number of foods. Make no mistake about it, from blue collar to haute cuisine, cheese is in. Cheeseburgers and pizza are the foods of teenagers, college students, and the working classes; cheese is also central to gourmet cooking.

Chic sandwich shops melt Swiss, Monterey Jack, cheddar, and a variety of other cheeses on top of everything from alfalfa sprouts to greasy pastrami. Mexican restaurants have mounds of cut-up cheese to sprinkle on everything and anything. Fancy Italian restaurants drown their pasta in sauces rich in cream, melted cheese, and olive oil. The breakfast croissant with an egg, already rich in fats and cholesterol, is now routinely supplemented with a slice or a slab of cheese.

If all of this sounds mind-boggling, it really is. Having laboriously decided (by the limited means available) what is bad for us in foods, we carefully package these evils in square slices, or chop them into small pieces and put them on top of virtually everything we eat. It seems hard to develop any rational explanation for such behavior. All of this has happened at the same time that we have been exposed to a flood of magazine and newspaper articles explaining the hazards of the big three: fatty acids, cholesterol, and salt.

There is a level of irrationality in our culinary behavior that seems especially difficult to analyze. There is not much in the way of sociological explanation for a nation's major culinary shifts. In the United States, we are very eclectic: We have incorporated the melting pot philosophy into our cooking pots but have nevertheless been swept by a cheese craze.

For those who have been sated by all this ferment, I recommend the following cuisines. Kosher meat restaurants absolutely will not serve cheese, under divine orders. Many Asians tend to eschew foods made out of fermented or rotten cream and milk. To my knowledge this includes Korean, Chinese, Japanese, Thai, and Vietnamese menus. I've never seen cheese served at a luau. There are also vegetarian cuisines that use no animal products. In that connection, my frequent attendance at college dining halls has led to the observation that most serve a vegetarian alternative at lunch and dinner. The main course in these meals is usually heavily laden with cheese. Thus, if the students are eating vegetarian to avoid saturated fatty acids, cholesterol, and salt, they are definitely not achieving their goals.

I have written about this cheese dilemma before, and the prose received little attention. I feel like a voice crying in the wilderness. No, not the wilderness! I am a voice crying in a lush verdant grassland, where my words are answered only by the lowing of cows, who contentedly chew their cud and participate in converting cellulose, indigestible by humans, into a variety of lipids all too readily assimilated and transported by humans to places where they may do us no good at all.

17

A Bioenergeticist's Revenge

YOU PROBABLY THINK IT'S EASY being a bioenergeticist, spending day after day interconverting kilocalories and kilojoules and trying to think about the reaction changes of entropy along the metabolic pathways. Well, I don't want to complain, but those of us working in the thermodynamic foundations of biology don't get a lot of respect.

Some time ago, I arrived at Berkeley to lecture just after the publication of my book *Foundations of Bioenergetics*. On hearing this news, an old friend replied, "Hey, I just saw a book on bioenergetics at Cody's Bookstore." Wishing to find out who the competition was, I hurried down to Telegraph Avenue to purchase the work in question, which was titled *Bioenergetics*. As I rushed through the pages to survey the new-found work, I realized to my shock and surprise that it was a how-to manual on sexual massage. As I say, there is not a lot of respect for those of us working in biological thermal physics!

Then things turned around, and I was asked to be an expert witness at the creation versus evolution trial in Little Rock, Arkansas. My area of expertise was "life and the second law of thermodynamics." After 30 years of having no one listen to me, I was the center of attention. When I came out of the courtroom, CBS, NBC, ABC, and the BBC would stick microphones in front of me to ask, "What do you think?" What I thought but did not say was: You reporters are not going to take the time, effort, and concentration to understand what I really think about this subject. I was on the Ted Koppel program, *Nightline,* and he also didn't take the time, effort, and concentration to understand what I thought. As noted, respect is hard to come by.

Shortly after the Little Rock trial, when I was still bathing in the limelight of my 15 minutes of fame, a new assault on bioenergetics began. Every time I picked up a newspaper or turned on the radio, the phrase "empty calories" came blaring forth. Empty indeed! They were referring to the stuff that undergoes glycolysis, revolves around the citric acid cycle, and funnels into the ultimate beauty of oxidative phosphorylation. This is the core of bioenergetics, part of the ineffable mystery of life. I spent a good portion of my professional career thinking about these matters, and in addition I have made a living at it. It may have been empty calories to them, but it was bread and butter to me.

I could never tell what was meant by empty calories. A calorie is a measure of energy, and energy is the *sine qua non* of life. It is to acquire energy for the biosphere that the leaves of plants reach out toward the sun in a phototropic response and branches grow upwards in a phototactic response. It is mainly to obtain energy that we eat. Energy is used to carry out our biological functions and to battle the entropic decay constantly breaking down our exquisite macromolecular structures. Empty calories, indeed—it is empty heads we are dealing with!

Well, all of this would have been part of the past but for the headline that leaped at me from the *Washington Post* as I was dawdling over my breakfast coffee. "U.S. Drops New Food Chart," it said, and the text was accompanied by a picture of "the eating right pyramid, a guide to daily food choices." The so-called pyramid has four levels. At the bottom, and the largest at six to 11 servings (I can't figure out what a serving means on this chart, but no matter), is Group One, a category labeled "bread, cereal, rice, and pasta group." Under the guise of a new name, the empty calories have returned, triumphant, as the very foundation of the eating right pyramid. " 'O frabjous day! Callooh! Callay!' He chortled in his joy."

One level up is the bifurcated botanical category consisting of the fruit and vegetable groups. The distinction is somewhat esoteric, and consulting my dictionary helps not at all. In any case, plant material other than grains are in Group Two except for some special cases.

The next level is subdivided into the milk, yogurt, and cheese group on the one hand and the enigmatic meat, poultry, fish, dry beans, eggs, and nuts group on the other. The latter is puzzling because one wonders why beans are Group Two and dry beans Group Three. If I pick fresh limas, my succotash will clearly be Group Two. If I munch on dry limas, I'm in Group Three. If I rehydrate the limas to make some more succotash, am I in Group Two or Three? It's a puzzlement!

At the top of the pyramid, with the ominous warning "Use sparingly," comes Group Four—fat, oils, and sweets. This is a very puzzling designation because "sweets" describes a flavor, not a food. The most common foods fitting this description contain sucrose and fructose, which should place them in Group One. What raises the rating of monomers and dimers of six-carbon sugars from Group One, which consists of high polymers of sugars, to Group Four? This classification is particularly strange because salivary amylase and intestinal invertase digest all of these materials into glucose and fructose, which are absorbed into the bloodstream. Many will recall an early school science experiment in which chewing on starch produced glucose, as indicated by a colorimetric reaction for sugar.

One detects an ideologic component to the Department of Agriculture's pyramid power. Materials at the bottom of the pyramid are viewed as good for you, and they become progressively more undesirable as one climbs the pyramid. This is the ultimate vindication of calories, empty or otherwise. Except that a puritanical strain enters the thinking: If the calories taste good, they become the dreaded sweets and move from Group One to Group Four, the ultimate residuum of empty calories. This puritanical expression by the Department of Agriculture must surely violate the First Amendment on separation of church and state. Come on, you folks in Beltsville, give us biochemistry and leave damnation to others!

This value-judgment-laden pyramidal representation is the reason for the article in my morning newspaper: "Yielding to pressure from the meat and dairy industries, the Department of Agriculture has abandoned its plans to turn the symbol of good nutrition ... to our 'eating right' pyramid." The meat and dairy industries had also sensed the puritanical nature of the pyramid and realized that their place in the triangle was next to that of fats and sweets, the least healthful foods. They were indeed worried that people might opt to eat less meat and dairy foods. And that is the advice of nutritionists. The political pressure worked, and the pyramid is being dropped.

So I am now left staring at a totem that was metaphysical in its conception and political in its abortion. Is that any way to advance public health? I'm also left wondering why oily peanut butter is Group Three and peanut butter oil is Group Four. And why is a triangle called a pyramid?

18

A Two-Coffee Culture

I WAS QUIETLY STUDYING my Linguaphone German book one evening and was startled to have Frau Pfaffinger state, "*Kaffee schadet meiner Gesundheit.*" The rough translation, "Coffee harms my health," reminded me that I am behind on my triennial analysis of the medical literature on coffee. I carry out this periodic ritual to prepare myself for the many discussions I seem to get into on coffee and the state of health.

The last time I conducted such a search, I sat with a big stack of *Index Medicus* volumes, looking up "coffee" in the alphabetical indexes and thumbing laboriously through the appropriate volumes. Well, technology has overtaken that procedure. I spent only a few minutes at a computer terminal in contact with MEDLARS, and one day later I was the proud possessor of a stack of $5\frac{1}{2} \times 8\frac{1}{2}$-inch printout pages containing 278 citations and ab-

stracts covering the findings on coffee reported by the worldwide
medical establishment from the beginning of 1985 to sometime
in 1987.

I would be less than faithful to my readers and indeed (Polo-
nius, please note) to myself if I had not gone through each of
those 278 entries, so I sat poised at my desk, reading away. I
must note with sadness that it is by and large a quite dismal
collection of material, for every statement occurs with its
negation:

Coffee raises blood pressure.

Coffee lowers blood pressure.

Coffee increases the incidence of pancreatic cancer.

Coffee has no effect on pancreatic cancer.

Coffee causes colon cancer.

Coffee protects against colon cancer.

As is well known from elementary logic, given a statement and
its negation, we can conclude anything we want.

In coffee research, as in all areas of study, fads seem to emerge.
A while back there was thought to be an association between
coffee and fibrocystic breasts. And now S. Hayday and J. G. Fodor
(*Canadian Journal of Surgery* 29:208, 1986) report: "On the
basis of their results and a review of available studies, the authors
believe there is no scientific basis for an association between the
consumption of methylxanthines and the development of fibrocystic breast disease."

The advantage of having all the abstracts before me at one time
is that it makes clear the overall quality of the information. Many
of the papers are epidemiological studies involving too few people
and too imprecise questions. The meaningless mélanges are then
covered over by statistical analyses to place cosmetic veneers on
flimsy results. I do not understand why investigators play these
games, but it sure leads to low signal-to-noise ratios when one
seeks to draw generalizations from the data. The medical profession's ability to offer reasonable advice on foods is continually
compromised by these large numbers of dubious and conflicting
studies.

I do not wish the previous statement to be a blanket indictment of all 278 entries. They did include some more substantial epidemiologic studies and a number of straightforward biochemical and cell physiology studies of coffee and its constituents.

Part of the problem in the questionnaire studies goes back to the basic question, What is coffee? To compare an Italian drinking three cups of espresso per day with an Englishman downing three cups of tepid tan milk hardly makes for precision. In my own investigations, I have had to establish a unique scale. When my daughter-in-law Jayne makes coffee, one cup is equal to 1.5 of my own, and when I visit California, I have determined that one cup of coffee brewed by daughter-in-law Miliani is equal to 2.6 of mine. So, for example, when one correlates coffee drinking with serum lipids, it is well to ask about how much cream goes in per cup, something not usually done. A cup of half-and-half per day could make a substantial difference.

Not wishing to be totally negative, I sought to salvage something from my latest search of the literature. There is, for example, confirmation that we have now become a two-coffee culture, as many of the researchers use coffee, decaffeinated coffee, or just caffeine to try to sort out the effect of the major alkaloid from that of other components. By and large, though, these studies suffer from the same deficiencies in population size and precision as the one-coffee investigations.

For whatever reason, it is clear that large numbers of people are opting for coffee without caffeine. The extent of that shift became clear to me when I walked into a local coffee and doughnut shop one morning and was greeted with the query, "Do you want decaffeinated or high-test?" And every major restaurant now offers both kinds of coffee. I have always been amused at the extent to which waiters and waitresses are real pushers when it comes to coffee. All other foods and drinks are sold in fixed measure, save coffee alone, the quantity of which is limited only by one's firmness in informing the server, "No more."

For those who think I am minimizing the hazards of coffee, I would like to report on two studies. The first, by R. Ruddy and associates (*Pediatric Emergency Care* 1:184, 1985), analyzed 250,000 furniture-related injuries in the United States in 1982 and found that fully 25% of them were inflicted by coffee tables.

Most of the victims were children; they sustained lacerations to the head and face from the sharp edges of square or rectangular tables.

The second study, by P. Lyngdorf (*Burns, Including Thermal Injuries* 12:250, 1986), addresses the epidemiology of scalds in small children, and it concludes, "Most accidents took place in the dining/living rooms and most were caused by coffee." These two papers require no sophisticated analysis to bring home the message that the best immediate public health advice on coffee is to buy only round or oval coffee tables and to drink it tepid. These simple expedients will dramatically reduce childhood trauma.

In any case, it is always possible to study the statistical correlation of coffee drinking with any and all of the "slings and arrows of outrageous fortune" that man is heir to. But unless the effects are acute and dramatic, the small sample sizes and large numbers of confounding factors render the studies quite meaningless, even though impressive statistical indices may be attached to the results. Common sense is a criterion that must be applied before statistical significance is considered. (This is not an attack on statistics, a vital tool in research. It is a comment on the abuse of statistics.)

So, in the end, I cannot tell you whether drinking coffee is good or bad for you. However, we note that, statistically speaking, your drinking coffee is bad for your young children, particularly if you put it on a low coffee table with sharp corners. And remember what an old Roman once said: "Moderation in all things."

19

Deliver Us

THIS ESSAY may get some people irritated with me, but I feel a need to write it. It is a nonprofessional's look at the practice of obstetrics. After 10 years of holding back, what finally prompted me to consider stepping in where angels fear to tread was an article in the *Washington Post* reporting that the rate of cesarean births in the District of Columbia is currently 30.1%. The national average is 24%, and one hospital in Maryland has reported a rate of 42%. The article also noted that "the chance of a woman dying during a cesarean section is two to four times greater than during a vaginal delivery, and the surgery involves a longer hospital stay, higher complication rates, and a longer recovery time."

The reasons most often cited for cesareans are fetal distress and failure of labor to progress. In the minds of obstetricians, who may be acting hastily, are threats of massive lawsuits if there are subsequent problems with the baby resulting from any delay. I might still have resisted writing this piece but for a coincidence. Shortly after reading the *Post* article, I found myself walking through Washington on my way to the Hungarian Embassy to get a visa. One of my purposes was to visit the medical museum devoted to Ignaz Semmelweis, a person who did so much to make childbirth safer. Thoughts of this noble and tragic man provided an impetus for this piece.

Before beginning my critique, I had best establish my meager credentials. Contrary to what a number of people think, and in spite of my having been on a medical school faculty, my advanced degrees are in physics and biophysics. With the exception of kidney surgery on a few octopuses, my experimental work on animals has been confined to organisms less than five millimeters in length.

My background in obstetrics has a rather idiosyncratic character. When our first child was imminent, I drove my wife to the hospital with some rapidity, and 10 minutes later an intern delivered my wife of a girl. The birth of our second child was a more complicated story. We had for various compelling reasons moved from Connecticut to Maryland in the seventh month of a pregnancy. My wife, who had been engaged in a natural childbirth prenatal program, located an obstetrician who professed an interest in the method and agreed to a delivery without anesthesia.

We arrived at Georgetown University Hospital in a timely fashion. One hour later the birth was proceeding uneventfully. The obstetrician, without consulting the fully conscious patient and without any medical rationale, administered a spinal anesthetic. Ten minutes later a son was born, and 12 hours after that the mother had excruciating back pain and was paralyzed in both legs. The physician insisted that it was no fault of his and the patient was malingering. I knew that the lady in question wouldn't know how to malinger if she had to. She came home by ambulance a week later, and it was six weeks before she got back the full use of her legs. That physician's actions and attitudes still enrage us when we think about them many years later. In a more litigious age, there would have been an investigation; at the time, we were naive, unaware of any recourse, and grateful that nature was taking care of this iatrogenic illness.

When the next pregnancy came along, the mother resolved not to get to the hospital in time for the obstetrician—a different one, to be sure—to do any unnecessary procedures. As a result of a slight delay, I was wakened in time to call the local rescue squad and minutes later assisted in the birth at home of No. 2 son. For reasons still unclear to me, we then went to the hospital to meet the doctor. Mother and child spent two restful days there.

For the fourth pregnancy, we were back in Connecticut, going to the same obstetrician who had missed the precipitous delivery of our first child. Although we respected his judgment in most matters, we could not agree to his plan to induce labor to ensure a conventional hospital delivery with a physician in attendance. We found a more flexible doctor, and a normal pregnancy ensued.

This time, the mother again planned to give birth a full six or eight minutes after arriving at the hospital, and I was summoned home from work in midafternoon. The timing calculations were again a few minutes off. When we arrived at the hospital, I rushed in and asked the staff to bring a stretcher for an imminent delivery. The nurses insisted I was panicking and finally arrived with a wheelchair just as I was helping in the birth of our fourth child on the back seat of our station wagon. At this point, two nurses kept running back and forth from the hospital to the car telling us a doctor would be there any minute. Mother, father, and child were doing fine. We finally persuaded the nurses to bring a pillow, a blanket, and something to wrap the baby in. In due time an obstetrician arrived for the delivery of the placenta.

Since the delivery had been performed under unsterile conditions, mother and child were then taken upstairs on a freight elevator with some workmen, including one who was smoking an outrageous cigar. This son of ours occasionally indulges in a cigar, and I still wonder if it is the result of this early imprinting. The two unsterile ones were isolated from the rest of the obstetric service, and early the next morning we decided that they should come home to avoid all that cigar smoke and other miasmas.

Yet another pregnancy, and we realized the futility of trying to time hospital arrivals with such precision. We persuaded the obstetrician—the same one who had missed the delivery for our fourth child—to agree to a home delivery. He, of course, had some conditions. First, if one of his other patients was giving birth at the hospital at the same time, that took precedence and we were on our own. Second, we had to agree not to tell anyone about our arrangement, as he definitely did not want to make a habit of home deliveries. To prepare for an emergency, I began to wear an umbilical cord clamp as a tie clasp.

At the appointed hour, the doctor arrived about 15 minutes before the birth, and I assisted in the delivery. The other four children awoke the next morning to greet their new brother.

Having established my meager credentials in the field of childbirth, I shall move on to a brief review of some obstetric practices of the past. In the mid-nineteenth century, a leading cause of maternal mortality was childbirth fever. We now know that in lying-in hospitals this was clearly an iatrogenic disease carried from dead and dying women to healthy ones on the hands of physicians. This connection was independently discovered by several investigators, but the most dramatic story is that of Semmelweis, who was driven to rage and despair by the failure of his colleagues to heed his plea for clean hands.

By the time aseptic procedures had finally ameliorated the scourge of infection, most women who could afford it were having their babies in hospitals. Good medical practice called for one to two weeks of bed rest after delivery. Many women rose from their hospital beds to become ill or die of thromboembolism. By the time death by extended inactivity was being eliminated, all the best obstetricians were taking prenatal X rays routinely and irradiating very radiosensitive fetuses with more roentgens than we can now possibly condone, given the slow film then available. The harm in terms of damaged babies and genetic load was probably very large.

The age of X rays overlapped, but was preceded and outlasted by, the age of overuse of obstetric forceps. Here again, one can only guess at the number of babies who were neurologically impaired as a direct result of over intervention by physicians.

Clearly, we are looking at the dark side of a noble profession that has accomplished much good. Nevertheless, this frightening historical record has to be dragged out from time to time to remind obstetricians of the need periodically to examine and reexamine their procedures. And when the rate of cesareans reaches 30%, this nonprofessional's built-in, waterproof, shockproof crap detector is telling him the time has come for obstetricians to take a very hard look at their paradigms.

IV

20

Roots of an American Tail

BEING A SCIENCE WRITER, I don't ordinarily comment about movies; in fact, I can't recall ever having done so. However, *An American Tail* comes very close to my "roots," and since some of the facts in the movie version don't agree with a brief oral history I have collected, it is hard for me to let the event pass without setting the record straight.

First, the chief character should have been Fievel Morowitz, not Fievel Mousekewitz. He came from Kapulya, not Shostka, and he emigrated to America in 1910, not in 1885. But, like the movie's hero, he did greet the Statue of Liberty with wide-eyed enthusiasm. You may ask, how do I know so much more about this man than the renowned Steven Spielberg? That's easy. Fievel Morowitz was my father.

For the first 40 years of my life, I thought Kapulya was a

mythical locale in western Russia, a symbol of the old country, a peg on which to hang some tales. This idea followed from the fact that I could never find it on maps or in encyclopedias. Then one day Dad casually let slip, with some pride, that Mendele Mokher-Seforim (Mendele, the seller of books) had been born in Kapulya. That led me to the *Great Soviet Encyclopedia,* where I discovered that Mendele Mokher-Seforim was the pen name of Sholom-Iakov Broide, or Abramovich, born in Kopyl' in 1836. I had never found "Kapulya" because I didn't know how to spell it.

Further reading of that encyclopedia informed me that Kopyl' is now an urban-type settlement and center of the Kopyl' Raion Minsk, Oblast Byelorussian SSR. It is six miles from the Timko-vichi railroad station and has a creamery and food-processing plant.

On subsequently learning more about Mendele Mokher-Seforim, I was dumbfounded that I had never heard of him, for he is almost universally regarded as the founder of modern Yiddish and Hebrew fictional writing and, additionally, penned a three-volume work on natural science. Note that the *Jewish Encyclopedia* (Funk & Wagnalls, 1916) spells his name as "Abramowitsch" and places Kopyl' in Lithuania. This confusion of places and names in transliteration makes it clear why "Speelburg" used "Mousekewitz" instead of "Morowitz."

The misplacing of Kopyl' is also easily understood, for the Minsk area was annexed by the Grand Duke Gedimin of Lithuania in 1321. Not until 1772, after a period of being under Polish rule, was all of Byelorussia (White Russia) ceded to Russia. When Mendele was born, Kopyl' was clearly part of czarist Russia.

Regardless of who ruled, the Jews of Kopyl' were clearly Litvaks and a blend of two streams of immigration. One came through Armenia in the first 200 years of the Common Era, and the second opted to move eastward to get out of the path of the German Crusades in the twelfth century. Litvaks are characterized by a great dedication to learning, an insatiable appetite for pickled herring, and an appreciation of an occasional nip of schnapps or slivovitz.

The Morowitz family emigrated to America not together but in several waves. The first wave was generated when my grandmother discovered her two oldest daughters, Chaia and Feitche, hiding in a large brick oven and sewing a red flag. Those two became the first to leave.

America was not entirely unknown to the family Morowitz. Lazar, my grandfather, was an itinerant carpenter who somehow traveled around the world on building projects and made enough money at that, and the family farm, to be regarded among the prosperous of Kopyl'. One of these trips had taken him as far as Jerusalem, and another terminated in Sioux City, Iowa.

After returning from America, Lazar and another traveler from Kapulya volunteered one evening to speak English for their friends. Years later, Fievel recalled only fragments of the conversation. One speaker began, "Goddamn it." The other replied, "Sure, Goddamn it." The rest of the English lesson is, alas, forgotten.

Lazar was not enthusiastic about moving the family and was indeed off on one of his foreign junkets when his wife, Chana, née Pessnachovich, saw red and made the decision alone. She was that kind of woman. And so, in 1908, the two daughters left.

Fievel emigrated next by himself. First came the six-mile horse carriage trip to Timkovichi, then the long train ride to Riga in Latvia, then the long, long boat trip to Ellis Island and New York. America, the land of opportunity, was where Fievel wanted to be. Years later, when he was 92 years old, he still wondered out loud with obvious disappointment why his father had not moved the family earlier so he could have gotten an education and become a professional.

The two things that were most surprising to Fievel on the trip over are a measure of how parochial life was in Kapulya. His first eye-opener was in Riga, where he saw a Jewish tailor working on the Sabbath. In 16 years in Kapulya, he had never encountered a Jew working on Saturday. The second surprise came when the boat stopped at Hamburg and he saw a black man for the first time. No one from Africa had ever come to Kapulya.

The young Fievel was sensitive to certain kinds of inequities, a sensitivity that never left him. Once when he was about 80, I asked him to recall for me something about Kapulya. He said, "One Saturday afternoon we were studying in the Synagogue. A fight broke out between two children. The father of one of them went over and beat the other child. The lad who was beaten was an orphan, and no one stood up for him. Yet we had just been reading in the Prophets about defending the widow and the orphan." That feeling for the disparity between word and deed never left Fievel; the image was still clear 70 years later. He would watch television and vocally chastise our political leaders about this disparity.

In any case, Fievel came to America and became Phil or Philip. Actually, he legally became Frank, and that's another American tale. You see, when at the earliest opportunity he went to take out his first citizenship papers, he did not speak English very well. He was accompanied by his brother-in-law, Abe, who had married Chaia, now Ida. Abe also did not speak English very well. The first question Fievel was asked was, "What is your name?" He replied, "Philip," and was met with a second question. "Do you spell that with one L or two L's?" The query clearly went beyond their knowledge of the subtleties of spelling in a second language. However, citizenship papers were too important to be compromised by linguistic uncertainties. Abe and Phil looked at each other, and Abe said, "Put down Frank." It worked, and Fievel legally became Frank, although everyone called him Phil for the next 70-odd years.

Why Fievel should become Philip is an interesting point in itself. Fievel is not a biblical name like Moishe or David; nor does it appear to be of Germanic origin, as is a name like Hirschel. My conjecture is that it derived from Philip or Philo, both names extant among the Jews since the reign of Philip of Macedon. One of Herod's sons was Philip, and one of the inscriptions in the Jewish catacombs of Rome of about the second century reads, "Here lies Eutychis, daughter of Philipus." A related name, Pfeyl of Passau, occurs as that of one of the 10 Jews brutally killed for "Host Desecration" in 1478. In any case, Feivels usually became Philips and sometimes Franks.

The rest of this American tale is the story of settling in, learning the language, making a living, and becoming a businessman, husband, father, and community leader. I am sure it is a scenario like a million others; indeed, the power of the movie version, released on the hundredth anniversary of the Statue of Liberty, is that it is a roots experience for many of us. The rest of the

Fievel Morowitz story is indeed the typical American experience, and only the end needs telling.

In 1980, a full 70 years after first seeing the New York skyline, Fievel (Phil-Frank) again emigrated, this time from New York State to Hawaii to spend his remaining years with his daughter and son-in-law. There he did volunteer work for several organizations and enrolled in history classes at Honolulu Community College. A lifelong ambition of attending college was being partially fulfilled.

I last saw him in the spring of 1985, when I was in Honolulu to give a talk on "Pantheism and Modern Science." Phil decided he wanted to come, and although he was having difficulty getting around, he made it to the auditorium and listened attentively. I was a bit concerned that I might offend his religious sensitivities and inquired about this after the lecture. He smiled and replied, "When you were a kid, you and Zayde [Lazar] used to argue about things like that." Indeed we did.

A few weeks before he died, Phil remembered there was something he wanted to do. He set up educational trusts for his great-grandchildren. As I said, Litvaks have a great dedication to learning.

21

Justice

It has taken me much of a lifetime finally to appreciate an anecdote I heard my father tell when I was 10. Dad was a businessman who had many lawyer stories, not in the modern hate-the-lawyers vein, but with a deep appreciation of the vagaries of the law. He once remarked that had his family come to the United States earlier, he probably would have become a lawyer. But, alas, the problems of earning a livelihood forced him to postpone his education until his mid-80s, when he enrolled in a community college.

My father's anecdote dealt with a man engaged in a civil action. When the decision was rendered, the man looked at the judge and pleaded: "Justice, justice!" The judge looked at him with some surprise and said, "Sir, this is a court of law, not a court of justice." The moral is that civil law is not a search for some vague absolute like justice but a pragmatic paradigm for resolving disputes.

If I had not been so starry-eyed, this arbitrary nature of the law should have been apparent from even a casual study of American history. For we have chosen to reify law by the decisions of nine justices, and often the vote is 5–4 or 6–3. Those split votes indicate the relativity of law. It is not a case of "anything goes," for there are rules to the game. Nevertheless, there is a great deal of latitude. So although we may piously speak of "a nation of laws, not men," laws are ultimately made tangible by the opinions of men and women. In the division of activities between the rhetorical and the metaphysical, law is found much closer to the rhetorical pole. There is much confusion, since the word "law" is used for three radically different constructs: judicial law, natural law, and divine law. Here, we are concerned with judicial. The other two are presumably less arbitrary.

Something has happened in recent years that has substantially altered the nature of civil law in the United States since my father's time. I would like to illustrate the present situation with a civil case I recently witnessed. The judge in this dispute had ordered a pretrial mediation session between the litigants. The court-appointed mediator was a retired judge. In addition to the mediator and the litigants, four attorneys were present, one representing each litigant and two appointed by the court in various capacities.

The mediator spoke slowly with a deep southern drawl. He also had a wealth of lawyer stories, Sam Ervin-style, which he related lengthily and deliberately. Given the cast of characters, these anecdotes were costing the litigants about $1,000 per hour, or $250 per story. Pretty pricey.

The mediator had only one technique of conflict resolution, the lose-lose game theory. He lectured on the high cost of litigation, which would eat up the entire amount under dispute in legal costs. He was, on the surface, totally unconcerned with equity. But he kept pointing out that if the case went to court, both sides would lose everything. An out-of-court settlement was reached.

The story illustrates that the cost of legal representation has gotten so expensive that in cases of less than several million dollars, legal fees become a dominant part of arriving at a solution. The paradigms of civil law have been so altered by the cost of lawyers and associated expenses that legal remedies are no longer available for most torts.

We can examine some of the consequences of the "gold-plated lawyer." First, and perhaps worst vis-à-vis legal structure,' every court case has four concerned parties rather than just two litigants. If the legal fees are an appreciable part of the settlement, then the interests of the attorneys become separate from those of their clients. For example, suppose that it is in the interest of the litigants to settle very quickly and that it is in the interest of the attorneys to prolong the dealings for many, many hours at X hundred dollars per hour. In such a case we can rely only on the honor of the attorneys to represent their clients. This is a lot to expect when the rewards are so high and the answers so iffy. Vision is often blurred through green-colored glasses.

One other feature emerges from the current situation. Every case proceeds by a kind of brinkmanship, a situation in which the least rational litigants are likely to win, for they are more willing to enter into the lose-lose mode than to compromise. The net result of this feature is to add even more arbitrariness to the proceedings.

Given these cost-driven problems with law, the first question that emerges is, Are lawyers overpaid? Coming from academics, a domain in which the superstars are paid less than second-rate journeyman attorneys, I incline to the position that lawyers are grossly overpaid. If we reimbursed our professors at the same rate that we reimburse lawyers, I estimate that college tuitions would be between $100,000 and $200,000 per year, and education would become a lose-lose proposition. What is there in the market place that drives up the cost of law? I pose this question to my colleagues in economics. Perhaps it should be directed to ethicists.

I do sense major conflicts of interest. Most legislators in the United States are lawyers. The people who make the laws are potential beneficiaries of a system that leads to inflated legal costs. Judges are also lawyers, so that the people who interpret the law and often set the fees also stand to benefit by a system that leads to inflated legal costs. If we wish to hold down the price of justice, there are severe structural problems in the system that must be dealt with first.

   In any case, redress of torts is being priced out of the market. The time seems at hand to reexamine our conduct of law in the United States. If justice is too idealistic a goal, society requires some sense of equity in order to discourage all procedures from becoming extralegal. My feeling is that this sense of equity is falling victim to avarice, and this is a substantial threat to a society of laws.

22

The Sidewalks of New York

I WAS WALKING DOWN SEVENTH AVENUE in a particularly sad and pensive mood, for I was returning from the funeral of two nephews whose lives had abruptly ended in an accident. Each intersection I passed from 39th Street to 33rd Street was the scene of some gridlock, which resulted from drivers moving into intersections when there was no clear route ahead. The obvious thought emerged that people thoughtlessly trying too hard to get an advantage were tying things up for themselves and everyone else. The clear and simple moral was noted, and I was searching for clear and simple morals.

At the entrance to Pennsylvania Station sat a one-legged man with a large scar across his face. He was repetitiously saying, "Vietnam veteran here. Won't you please help?" I averted my eyes and walked down the stairs. At the bottom I turned, went up the escalator, and handed him a dollar bill. We looked into each other's eyes, and he said, "God bless you." Upon arriving at the information board I was approached by a man who said, "When you're homeless, it's hard to get any sleep—the police keep moving you." I asked him about going to a shelter, and he replied, "At shelters you're mistreated." "Look," I said, "I can't solve your housing problem, but take this." Again a dollar bill. We also made eye contact, and he took the bill and said, "You know, that's like a million dollars to me. I want you to know I'm going to get some hot soup."

I must confess that I am not usually given to passing out money on the byways of New York. My last attack of giving came in Nairobi, where I at first toured the streets averting my eyes from the limbless and sightless who sat pleading. Then one night I was cogitating while eating a superb vegetarian dinner at an Indian restaurant. The main dish cost 100 shillings (about $6). I began to think how much 100 shillings would mean to the beggars of Nairobi. How, I thought, could I ever accomplish so much for so little? From then on, I began each day with a pocketful of coins and neither avoided eye contact nor was deaf to any voices as I wandered the city streets distributing shillings and trying to understand a different world.

I stood in Penn Station thinking of the homeless, a subject that greatly troubles me of late. It is a cliché to say that it is one of our great national failures, but like most clichés, it is also true. What resources, I thought, would be required to solve the problem of our street people?

Once seated on the train, I took out pencil and paper. I recalled reading that the Savings and Loan and HUD thefts would cost the taxpayers at least $210 billion. If half of that money went to build $50,000 housing units, we could have 2.1 million such units, more than are needed to house all the homeless. If the other half went into a trust fund, we could have $8 billion a year in interest to keep up the housing and help the needy. If the rentals on the units averaged a mere $100 a month, we would have an additional $2.5 billion a year for ongoing construction. Clearly, there is no shortage of money to solve the housing problem. There is a shortage of will. The funds that could have easily solved the problem will not go to the needy, because they have been ripped off by the greedy. The villains are light-fingered gentry, land sharks, brigands, white-collar thieves. Our national resources have gone into a welfare fund for a small group of the immoral rich. (Do you hear me, Mr. Watt?)

And how has this happened? Could the federal government have kept better track of this money? Of course it could have, as is demonstrated by another personal anecdote.

A year ago I was undergoing one of the less pleasurable duties of citizenship, a routine tax audit. At one point my auditor, a pleasant and competent young person, raised an issue of $14.52. For sentimental reasons I maintain a savings account in Hawaii, which had earned for the year in question the less-than-grandiose sum just mentioned. I had entered the amount on Schedule B. However, the serial number on the passbook I had dutifully produced did not agree with the account number on the Form 1099 in the agent's file. The matter was resolved by my finding that the number on the paper passbook holder was in error and the digits in agreement with the 1099 were indeed inside the passbook. I present this story to demonstrate that a government that can keep track of the correct filing of a $14.52 entry clearly possesses the technical wherewithal to monitor its billions of dollars.

No, the failure is not arithmetic—it is moral. The greed epidemic is as rampant as the AIDS epidemic and perhaps as threatening to our national well-being. It comes from the well-to-do, the well-educated products of prestigious postsecondary schools. It spreads through the population. It leads some to excoriate welfare cheaters while skimming funds designated for the public good. (Are you still listening, Mr. Watt?) It is rampant in the investment scandals. I could not avoid thinking of what Yale graduate Edwin Meese symbolizes when I recently saw a mobile soup kitchen a block down Vanderbilt Avenue from the New York City Yale Club.

Success is not a vice. But greed that achieves success at the cost of violating the rules of the game and that is uncaring about the less fortunate is not acceptable. We cannot survive as a nation if housing is beyond the means of our poorest citizens, who are forced to live on the streets.

As a lifelong teacher at major American universities, I keep asking myself, Where have we gone wrong in graduating so many obsessed with greed and self-centeredness and devoid of the sense of ethics necessary for a democratic society to flourish? Should our curricula have included more emphasis on the good as well as the true and the beautiful? Could we the faculty have been better role models for the young in impressing them with social values as well as individual wants? I do not know the answers to these questions, but I know that the battle against greed is our common responsibility. I shall make it a habit to contribute to the homeless. I do this in memoriam of my nephews Jonathan and Daron Regunberg.

23

Brighten the Corner
Where You Are

A WHILE AGO I was scheduled to take a very early morning flight out of Newark Airport. Since there was no easy way to get to the terminal from New Haven at that hour of the day, I opted to spend the night in New York at the Yale Club. The location somehow reminded me that I hadn't seen one of my undergraduate roommates for some time, so I called and we arranged to meet for dinner at the club. We formerly would see each other every couple of years, but somehow the periods had spaced out over time.

A part of the undergraduate experience that I recall vividly and fondly consisted of endless hours of bull sessions, in which the participants explored life and literature, ethos and ethics, science and sex. Part of the alumnus experience is the ability instantly to resume conversations with those who have been party to sufficiently long and meaningful talk sessions as undergraduates.

Indeed, before the waiter set a glass in front of us, we were deep in conversation. Jay had just retired after a career in New York City government, and I had recently completed my term as Master of Yale's Pierson College.

Five years before, when I had first taken over my duties at the college, I would occasionally call on Jay as a special consultant. On my arrival at Pierson, I had been told that the college had an uncomfortable reputation as a white preppy institution unfriendly to minorities. Jay, having grown up as a streetwise black in New York City, maintained contact with the current generation of young people. I found his advice helpful, and I hope I left the college with the students' having the feeling that Pierson was truly home to all of its residents.

When Jay J. Swift arrived at Yale in the spring of 1945, he was, I believe, the only black among the civilian Yale undergraduates. Another four or five blacks were enrolled in the Navy V-12 programs. In New York I joked with my ex-roommate that there are now more black cheerleaders at Yale than there were total blacks enrolled in the early 1940s.

Although he was hardly an ideologue, Jay was a pioneer, as the following story will illustrate. One Friday night four of us were out having some beers at George and Harry's on Wall Street. It was getting rough, as it frequently did on weekends, and one of our foursome who was a member of Mory's suggested that we go there to continue our beer and conversation. We arrived and sat down at an empty table. The manager came over and asked the member if he could speak to him in private. The message that was conveyed was that we were to leave because blacks were not allowed at the club. The four undergraduates departed, sadder and wiser in the ways of the world.

In a thousand ways it was difficult to be a member of such a tiny minority in a bastion of strong traditions, both good and bad. Whether it was where to get a haircut or how to interact with girls at social events, there were always special questions. Jay was good-humored, outgoing, and gregarious. He was a realist who had a talent for avoiding self-pity. He emerged from the Yale experience with a number of good friends.

Well, in this regard times have changed for the better. We have come a long way. That is said in spite of the obvious: We have a long way to go. I don't think it wise to focus only on bad news.

The conversation that evening turned to Jay's search for his roots, and he talked about his trips to the West Indies to seek out relatives and ancestors. His ties to Africa had been lost in history, and he would have liked to have known more about them. Jay was deeply concerned about the problems that many blacks were having in making it in American society. There were no easy answers, as we both knew, but exploring big questions had always been a major component of our friendship.

As we talked on about past and future, Jay asked me what it had been like being Master of Pierson College. I recalled the many sophomoric sessions when we had talked about changing the world. As Master, one had to give up global thoughts and focus on a much more limited domain. But in that small arena it was actually possible to do some things that had a major impact on some very fine young people. I had given up trying to change the world in return for working on just one small part of it.

Those thoughts reminded Jay of a hymn he used to sing in church as a child. He recalled the title: "Brighten the Corner Where You Are." I asked him if he could get a copy, and he said he'd try. We talked for a long time, and then he left to go home.

Two weeks later a photocopy of Hymn 65, "Brighten the Corner Where You Are," arrived with a note from Jay. I was delighted and want to share the words with you:

> *Do not wait until some deed of greatness you may do,*
> *Do not wait to shed your light afar,*
> *To the many duties ever near you now be true,*
> *Brighten the corner where you are.*

> REFRAIN:
> *Brighten the corner where you are!*
> *Brighten the corner where you are!*
> *Someone far from harbor you may guide across the bar,*
> *Brighten the corner where you are.*

> *Just above are clouded skies that you may help to clear,*
> *Let not narrow self your way debar,*
> *Tho' into one heart alone may fall your song of cheer,*
> *Brighten the corner where you are.*
> REFRAIN

Here for all your talent you may surely find a need,
Here reflect the Bright and Morning Star,
Even from your humble hand the bread of life may feed,
Brighten the corner where you are.
REFRAIN

Forty years after graduation, Jay and I were both naive enough to resonate to those words of Ina Duley Ogdon and Charles H. Gabriel. That was a happy thought.

About a year or so after our dinner, I was back in New Haven after a year on sabbatical. At graduation a mutual friend sought me out to tell me that Jay was in the hospital, very ill with cancer. At the first opportunity, I went to Roosevelt Hospital in New York to visit my friend and have what turned out to be our final conversation.

He was flat on his back, almost totally paralyzed. With amazing ease we resumed our previous conversation. We talked of the economics and politics of East Africa. He then asked me what I knew about chemotherapy, and we briefly reviewed the little I could recall. We talked of the past and discussed some mutual acquaintances. I sensed that this might be our last meeting, but Jay seemed quite optimistic. The conversation was a mixture of past, present, and future. It never admitted the ultimate: It was in truth a bull session.

A few days later, Jay died. I sit here at my desk thinking about how I might honor the memory of my friend. If I close my eyes and listen with my heart, I can hear the answer. A 1930s black congregation is belting out a song telling me to:

Brighten the corner where I am.

24

Nixon and Thermodynamics

RICHARD MILHOUS NIXON is not my favorite U.S. President, nor is he my least favorite. He did play a major role in determining my career, however, and I think that the time has come to acknowledge his contribution. I was about to say, "to give the devil his due," but that would be a more judgmental statement than I wish to make. Let the facts speak for themselves.

The story begins in the spring of 1951, when I was completing my Ph.D. thesis and looking for a job. With a wife, one and a half children, and a very unruly dog named Dodger, the search for employment was serious. Through the old boy network, my thesis adviser had learned that the National Bureau of Standards was looking for a biophysicist, and I was recommended for the job.

The director of the Bureau of Standards at the time was Edward

U. Condon, one of the country's most distinguished physicists. Condon decided that in order for the bureau to fulfill its standards mission, the staff should show competence in a number of areas that until then had been outside the bureau's domain. One of those areas was biophysics; he envisioned a small group that would carry out independent research and be available to help whenever biological issues came within the scope of the bureau's activities. Although some work on medical instrumentation was in progress, I was the first full-time biophysicist hired for the new group, which for somewhat mysterious reasons was administered as part of the Temperature Measurement Section of the Heat and Power Division. Of course, the section did calibrate clinical thermometers, but that hardly seemed a sufficient reason to make it the center of biophysics. Of more direct relevance was the fact that within the division was Herbert Broida, a molecular spectroscopist who was collaborating with heart surgeons at Georgetown University Hospital on the design and construction of an artificial aortic valve. To the best of my knowledge, the first such plastic valve used in a human was fabricated in the machine shops of the Bureau of Standards.

Late in June 1951, I arrived in Washington to commence my somewhat ill-defined duties as the bureau's biophysicist. I was given an office and assigned a laboratory in the Dynamometer Building. The laboratory was on a hallway off an enormous room full of extraordinarily elaborate machinery. Some workmen there were constructing large tanks. I inquired of the man in charge about the intended use of the tanks. He replied, "Outboard motor testing." "What kind of testing?" I asked meekly. He shot back, "Endurance testing." My days at the bureau were disquieting in more ways than one.

The first major purchase for my laboratory was an autoclave, because one of the planned projects involved the culturing of microorganisms. Shortly after the device arrived but before I had a chance to use it, I returned to my lab from the quiet of the library. A representative from the Property Inventory Section was placing a metal identification tag on the device. To accomplish this, he had drilled two holes in the cover plate and screwed on the tag. When I turned the autoclave on and it warmed up, two jets of steam issued forth from the screw holes. The problem was solved by taking out the screws and filling the holes with solder. I tell this story now that the statute of limitations frees me from concern as to which law I violated by removing the identification from this government property.

I arrived in the nation's capital in the midst of Senator Joseph McCarthy's attack on the federal establishment as a hotbed of Communism. A number of the nation's scientists were summoned to Washington to be pilloried by McCarthy and his associates. As a sideline to this lamentable scenario, sometime before my arrival a subcommittee of the House Un-American Activities Committee led by then Congressman Nixon had accused Condon of being "the weakest link in America's atomic security." Indeed, the Bureau of Standards had an ordnance laboratory heavily involved in military work. Shortly after my arrival, Condon decided that enough was enough and resigned his post to become research director at the Corning Glass Works. Word around the bureau had it that the then Senator Nixon was putting on the pressure. We feted Condon with a banquet, where he gave a speech announcing that for the past 20 years he had been a card-carrying member of the... Democratic party. A pall of fear and trembling hung over the bureau. Condon was replaced by Allen Astin, a well-respected scientist from the Ordnance Division.

From my noise-filled laboratory in the Dynamometer Building I observed the events in relative calm. I was free from sin, more or less. Nevertheless, when Astin reviewed the bureau's program, he saw no reason why the institution should be in the biophysics business. Since I was a civil servant, no one made any effort to fire me, but the funds paying for my research abruptly disappeared. It took no insight to see the handwriting on the wall. I began to look for another job. I also shifted my research to make use of the available resources, which included a superb physical sciences library, whatever supplies could be scrounged, and a laboratory well suited to study the effects of excess noise on human sanity. The Heat and Power Division provided me with pencil and paper.

I began in earnest to study the thermodynamic foundations of biology, a subject that seemed both fitted to my interests and appropriate to my surroundings. I was also in day-to-day contact with some physicists who possessed a deep understanding of thermodynamics and statistical mechanics. I began to think about Erwin Schrödinger's book *What Is Life?*, which raised important issues about living systems and the second law of thermodynamics.

In 1951 and 1952, I heard two talks, one by Norbert Wiener and the second by Leon Brillouin, that set me to thinking about information, entropy, and the relation of life to its thermodynamic foundations. I undertook a study of thermal physics and began to understand some constructs that I had had too little time to think about as a student. In short, Senator Nixon had presented me with the gift of time, I was in a rich intellectual environment, and I was driven by twin desires of succeeding as a scientist and understanding more about the underlying nature of life. I finally got another job and left the bureau.

Many of my subsequent activities as a scientist stem from a continuing interest in the thermodynamic foundations of biology as they relate to the question, What is life? I have no way of knowing what path my career might have taken if Nixon had not attacked Condon with such vigor. With sufficient funds for laboratory research there would have been far less incentive to study theory.

Occasionally, students come to ask my advice about subtle career choices. I find it hard to be wise or even overly serious, since I carry around the intimate personal knowledge of the kinds of things that can make a big difference in the unfolding of one's professional life.

25

"George Mason, One of Our Really Great Men"

THOSE BYLINE READERS among you will note that my niche has changed from Molecular Biophysics and Biochemistry at Yale University to Biology and Natural Philosophy at George Mason University. The relocation of one professor is hardly a matter of note, but somehow I cannot resist seeking for the deeper significance of each minor event in life. And this inquiry has, as is often the case, opened up a new stream of thought quite unrelated to the immediate act.

When communication first began with my new university, I quite naturally began to think about the George Mason for whom the institution was named. During my previous university association I had never looked deeply into the life of Elihu Yale. (Something in his past involving irregular transactions while a British East India Company agent in Madras has discouraged subsequent generations of Elis from concentrating on the life of the man whose philanthropic gifts, arranged by Cotton Mather, gained him a kind of immortality. As a twentieth-century Yale president is alleged to have said, "The only trouble with tainted money is there taint enough of it.")

In any case, early in my dealings with George Mason University I took down the encyclopedia volume *Krasnokamsk to Menadra* and there found an article noting that George Mason (1725–92) was an American patriot, statesman, and member of the Federal Constitutional Convention who did not sign the United States Constitution because it lacked a bill of rights and permitted continuation of the slave trade. His bill of rights in the Virginia Constitution is the model for the first 10 amendments to the United States Constitution. It seemed shocking that after all my years of reading American history I was totally unfamiliar with this man.

I went back to the two works on the United States that have always been canonical for me. Woodrow Wilson's five-volume *A History of the American People* has two passing references to Mason. Samuel Eliot Morison, in *The Oxford History of the American People*, accuses Mason and others of abstaining from signing the Constitution "largely from wounded vanity, since their pet projects were not adopted." Looking further, I perused a paradigmatic history text: *The Great Republic: A History of the American People* (Little, Brown, 1977). Written by six professors—two at Harvard, one at Yale, two at Brown, and one at Northwestern—this 1,267-page volume mentions Mason only once in a list of influential delegates to the Constitutional Convention. This, I remind you, is the historians' treatment of a man responsible for the Bill of Rights and much more. Scholars of American history have, de facto, written George Mason out of history.

I have now had a chance to go beyond the standard texts and have read *George Mason, Constitutionalist* by Helen Hill (Harvard University Press, 1938) and *George Mason's Part in Framing the Constitution of the United States and the Bill of Rights*, privately published by Lee Fleming Reese in 1988. I present a brief synopsis of what every American should know about this far-seeing statesman.

Mason spent his life as a northern Virginia farmer, increasing the very substantial estate left to him by his father. Largely self-educated and privately tutored, he seems to have received much of his education from his uncle's scholarly 1,500-book library. Mason married Anne Eilbeck, and they had nine children. A close friend and sometime political adversary of George Washington, Mason was a member of that remarkable group of northern Virginians active in gaining independence and leading the 13 colonies to becoming a constitutional democracy.

Mason suffered most of his life from painful "gout" and "colic," which limited his public life. His wife died in 1773, and during the period of his greatest political activity he devoted much time and attention to his children, whose ages at the time of his spouse's death were three through 20.

His public writings, beginning in 1765, oppose the further importation of slaves and promote the rights of man. He pointed out that the institution of slavery had brought about the decay and destruction of the Roman Empire. From the mid-1760s on, Mason's pen was constantly active in opposing taxation of the colonies by the Crown. When things began to heat up, he authored the Fairfax Resolves stating the liberties due to Americans.

From 1774 to 1776 he was engaged in establishing a government for Virginia that was independent of royal authority. At a later time, Thomas Jefferson noted in a letter (April 13, 1825) that "the Bill of Rights and the Constitution of Virginia were drawn originally by George Mason, one of our really great men." During the prerevolutionary period, Mason and Washington were active in organizing and provisioning a Fairfax militia.

The Virginia Declaration of Rights to which Jefferson refers was used in formulating the Declaration of Independence. Mason's first draft began:

> I. That all men are created equally free and independent, and have certain inherent natural rights, of which they cannot, by any compact, deprive or divest their posterity; among which are the enjoyment of life and liberty, with the means of acquiring and possessing property, and pursuing and obtaining happiness and safety.

Reading these words makes clear our debt to this Virginian. Consistent with an interest in rights, he aided in the preparation of

"A Bill to Prevent the Importation of Slaves" (1777) and "A Bill for Establishing Religious Freedom" (1778).

In 1787, Mason, Washington, James McClurg, Edmund Randolph, John Blair, James Madison, and George Wythe were selected as Virginia's delegates to the Federal Constitutional Convention. In the spring of 1787, Mason traveled to Philadelphia for the opening of deliberations.

The story of that convention is well known from the pen of Madison. Mason addressed the convention on almost every issue that came before it. As the long hot summer wore on, slavery became the central focus. To many delegates it was a point of contention between northern and southern economic interests. Mason headed a group who, according to Hill, "wished to treat slavery as a social institution rather than as the raw materials of bargaining power."

On August 22, the gentleman from Virginia made an impassioned speech to the convention on the evils of slavery. With amazing insight he foresaw the Civil War: "By an inevitable chain of causes and effects Providence punishes national sins, by national calamities." On August 25, by a vote of seven states to four, the continuance of the slave trade was passed.

When the final Constitution was complete, Mason could not in good conscience append his signature. He also opposed its acceptance by the Virginia Convention. On June 26, 1788, that body accepted the new document by a vote of 89 to 79. During the intervening discussions in 1787 he had set forth his objections in a closely reasoned document. It includes the statement, "There is no declaration of any kind, for preserving the liberty of the press, or the trial by jury in civil causes." His objections conclude with the point, "The general legislature is restrained from prohibiting further importation of slaves for twenty odd years."

The first Congress after the adoption of the Constitution passed the Bill of Rights, and on December 15, 1791, its ratification by Virginia, the last of the required states to do so, made it the law of the land. Part of Mason's dream for his country had come to pass.

When Mason lay dying, he was visited by his friend and neighbor Jefferson. Mason confided that even without the inclusion of a bill of rights he would have signed the Constitution, but he could not go along with the continued importation of slaves.

Now that I know who George Mason was and what he did, I must raise the issue of why the author of the Bill of Rights has been written out of American history. It has been said that history is the story of victors, and Mason was a loser both at Philadelphia and at Richmond. But I would argue that it is the business of faithful historians to record those who lose as well as those who win, especially when one of the former was motivated by concepts of such overwhelming importance in our constitutional system and ideals of such nobility. You may think that this is a severe criticism to make of some historians; well, it is.

V

26

Koobi Fora

MOST OF US like to be present at "firsts" or to visit sites where the original something or other took place. These firsts come with varying degrees of emotional impact. Thus, for example, a few weeks ago in Nairobi I was present at what was billed as the first performance on the African continent of Sonata No. 15 by Niccolò Paganini. Doubtless for Paganini devotees in the audience this was a magical moment, but to me it was merely an interesting footnote to a fine concert by Fernando Antonelli on the violin and Francesco Beroglio on the guitar. I was frankly more taken by the oft-performed Sonata No. 3 in C Major, with the extraordinary *Moto perpetuo,* in which the violinist plays 2,100 notes in three minutes.

The visits to sites of first events that have had the deepest impact on me over the years have been: arriving on top of the

mountain in the Sinai desert where humankind first came face
to face with ethical monotheism, and standing at the site of Socrates' prison in Athens, where humankind first came to grips
with a rational humanism. I now am off to experience another
one of these major firsts: Koobi Fora, the site where humankind,
in terms of the genus Homo, may have been born.

A few days ago in the office of Richard Leakey in the Kenya
National Museum, we had a brief discussion about new information on mitochondrial DNA and its bearing on our view of the
transition from *Homo erectus* to *Homo sapiens,* which is believed
to have occurred some 200,000 years ago. What one calls the
birth of humankind is purely metaphorical; it is difficult to say
with precision which nucleotide change in what individual constituted the emergence of humanness during a three-million-year
continuum. Nevertheless, a series of molecular events parallels
our most poetic descriptions.

Leakey, the director of the museum, kindly introduced me to
Emma Mbua, curator of hominid fossils, who admitted me to the
vault to see the originals of some of the skulls of Proconsul,
Australopithecus, *Homo habilis, Homo erectus,* and early *Homo
sapiens*. There was, I must confess, a strange feeling about being
in the physical presence of our earliest ancestors. Ms. Mbua treats
the collection with great care, mindful of the priceless treasure
it represents. She is one of a group of scientists with a special
mission: to explore where we came from. Of particular interest
was specimen 1,470, the skull of *Homo habilis* from Koobi Fora,
which places our earliest ancestors of the genus Homo in Africa
some 1.8 million years ago. I had read of these finds from northern
Kenya and left the museum with a firm resolve to visit the site.
It was one of those searches for firsts that cannot be denied.

Koobi Fora is in Kenya's most remote national park, Sibiloi, on
the northeastern shore of Lake Turkana (formerly known as Lake
Rudolf or the Jade Sea). The lake is a large landlocked body of
water some 160 miles in length and as much as 30 miles across.
It is in the northwestern part of the country. The northern tip of
the lake is near a point where the Kenyan, Ethiopian, and Su-
danese borders come together. The lake is about 100 miles east
of the Ugandan border. The geological history of this region enters into understanding the formation of fossils near the present-
day shores.

Resolving to visit and actually getting to Koobi Fora are two different things, and I spent the next day trekking the streets of Nairobi from safari office to safari office, finally locating Ivory Safaris, which books boat trips to Koobi Fora from Lake Turkana Lodge on the western shore. Two days later, I left Wilson Air Terminal in a five-passenger Kenya Airlines plane, which flew northeast around Mount Kenya and then northwest, making its first landing on a small airstrip at Loiyangolani Oasis. Two passengers got off, one got on, and we flew off across the lake and landed at an airstrip in the middle of the desert. It bore the name Kalokol. One passenger disembarked. I realized I was off the beaten path.

A Land Rover was waiting, a man with whom I did not share a language took my carry-on, and we began bounding off across the desert. About 10 miles later we came to the lodge, a cluster of 18 cabins facing the shores of the lake. The bar serves cold drinks, the small swimming pool is remarkably cool, because of the low humidity and high heat of vaporization of water, and the cabins are quite cheery. I have a slight caveat for those who may have a phobia for flies, 100° temperatures, small lizards in the cabin, and crocodiles and hippopotamuses in the lake. However, if you want to get away from it all, really hear the sound of silence at night, observe birds in large numbers, fish for 100-pound Nile perch, and see the skies in all their glory, I can recommend no better place.

I went to bed early. There is not a lot of night life at Lake Turkana, and the generator goes off at 11 o'clock. Up at five and out to see the stars. The inverted Big Dipper is pointing to the Pole Star, lost in the haze of the horizon (we are a few degrees north of the equator). A 180-degree turn and the Southern Cross looms high in the sky. Our coordinates are set, and an hour later the sun rises on the perpendicular to the north-south line. The morning is spent in preparation for our trip, and we eat an early lunch. As there are no other tourists on this safari, it is my private expedition. Indeed, as I watch eight stalwart Turkana men in a row carrying our supplies down to the boat, I feel like Bwana Harold, the last of the great African adventurers.

Our vessel, a 28-foot launch with two outboard engines, is covered with a sunroof. We take off toward Central Island after being bid farewell by Mr. Hasham, the lodge manager. The safari staff consists of Alila, the captain; Gabriel, his crew; Etir, the cook; and Peter Saina, the operations manager of the lodge. I am getting a bit light-headed in the noonday sun, and phrases like "Dr. Livingstone, I presume" echo in my mind.

The crossing to Alia Bay is fairly rough, and after three and a half hours we drift into the muddy shore and put out an anchor. At Alia Bay there is a park ranger station, and a Land Rover awaits us. We hurry off on park roads to get to Koobi Fora by dark. Along the way I see the first gerenuks and Somali ostriches that I have ever encountered. In addition there are Grant's gazelles, zebras, jackals, and hares.

At the Koobi Fora camp and headquarters we are greeted by Francis Kasuli Lili of the Kenya National Museum. He worked with Louis and Mary Leakey at Olduvai Gorge in Tanzania, and since 1976 he has been at Koobi Fora working with Richard Leakey. The man has enormous enthusiasm for his work, which shows in the warmth of his greeting for those who make the long trip to Koobi Fora to see what has been accomplished here.

Etir does his thing, and we settle down to dinner and off to an early sleep. The cabins have comfortable beds but are rather more open to the local fauna than I would have guessed. However, the only animals I encounter that night are a bat that flies through the cabin at around midnight and a lizard that I rescue from the wash bucket in the morning. One occasionally hears the roar of Simba the lion, but that seems quite far away.

At sunup Francis and I go off to the museum about a kilometer away for my preprandial lesson on fossil formation and the archeology of stone tools. A number of geological concurrences— the rising and falling of the lake level, sediment formation, the movement of the earth due to the formation of the Rift Valley, and volcanic action—produced conditions under which large numbers of fossils were formed during the past three million years and now lie on or close to the surface.

Here, some 1.8 million years ago, lived *Homo habilis,* presumably our lineal ancestor, together with *Australopithecus bosei,* a large-jawed hominid whose teeth indicate it ate fruit and nuts. Autralopithecus became extinct for unknown reasons, and *Homo habilis* presumably gave rise to *Homo erectus.* Although various stone artifacts indicate Australopithecus was a tool user, only with the genus Homo do we find evidence of a toolmaker. Near the Homo fossils are found workshops with both stone tools and the chips produced as the implements were made. The fully opposable thumb as well as the larger cranial capacity seem associated with the emergence of Homo the toolmaker.

We return for breakfast and then examine a number of fossils that are being cleaned up by very patiently chipping and scraping away at the sedimentary materials in which they are embedded. I begin to get a feel for the kind of pattern recognition that allows considerable information to be deduced from collections of bones and teeth. The amount of fossil material around here is huge, and much remains to be uncovered. The time comes to say goodbye to Francis, and we take off to visit three sites where major, almost complete vertebrate fossils are embedded in the rock. There is a giant tortoise, a very large crocodile, and an elephant-like animal with enormous tusks curving out on each side. At a series of outcroppings, mollusk shells are lying about or sticking out of the rock.

Just outside Alia Bay on a hillside is a petrified forest. Sections of large trees at one time fell into the water and became mineralized. What is strange is that these sections look as though they had been cut by a large saw. Since the logs date back to the time when *Homo erectus* roamed this area, one wonders what kinds of implements had been fashioned that could cut through large logs. Certainly, none of the stone utensils that have been found to date is capable of making such cuts. Another explanation is that the stone matrix that fills these ancient casts tends to split in transverse fracture planes when the logs are disturbed.

At Alia Bay we return to the boat, push it off the mud, and head across the lake. Three hours later we are back at the lodge. Sibiloi Park is the hottest, driest place I have ever visited, and I head for a cold drink and a dip in the pool. I have a feeling of tranquility, a sense of having just had the ultimate "Roots" experience. For doubtless, somewhere near here in East Africa, our humanity emerged from our primateness. To see the bones and touch the tools affirm that emergence. For to cherish the roots is to value the shoots, and they, dear reader, are you and I.

27

Clitoridectomy

MY EARLIEST KNOWLEDGE of East Africa came from a childhood hobby of stamp collecting and an attraction to the strange-sounding names. Somewhere I have tucked away an album with a few stamps from Kenya, Uganda, and Tanganyika. These three states were under British government and became independent in the early 1960s. They have traveled radically different routes since then. Kenya, under Jomo Kenyatta and his successor, Daniel Moi, has become a successful, Western-style, free-enterprise state. Tanzania (the former Tanganyika and the island of Zanzibar), under Julius Nyerere, has followed a Chinese style of communism, and Uganda, under a series of leaders, has followed a path of intertribal strife and chaos.

I have not come to Africa as a student of politics, however. Rather, I am here as a traveler to see the land, the people, and the animals and to trek up Mount Kenya. A few days' stay in Nairobi gives me a chance to set down some thoughts and to sense the tempo of the city. I have been walking about the streets, and as evening descends I stop at a place called The Pub to quench a thirst brought on by a hot late-afternoon sun. At the bar I meet a Tanzania-born Englishman who received his medical education at Cambridge and then returned to practice in Africa. I think: "What a wonderful chance to learn of East African medicine!" But alas, he gave up practicing for drinking a few years back and is hardly current or, indeed, all that coherent.

When the doctor and his bar companion, who I believe is an accountant, hear that I am newly arrived, they suggest that I accompany them for a drink at their next stop, Buffalo Bill's Wild West Saloon and Eatery. I shortly find myself at that establishment's bar next to an American who teaches at Kenyatta University. While we are engaged in professorial shop talk of this and that, my original companions leave, apparently headed for their next watering hole. The professor's Ugandan wife arrives, and we chat awhile about their visit to the United States and then move to a booth with some other folks standing at the bar. The soft-spoken woman on my right introduces herself as Victoria. She is also from Uganda. How appropriate a name, I think, for someone coming from a country bordering Lake Victoria. Scenes from my all-time favorite motion picture, *The African Queen*, begin to flash before my mind's eye.

As the conversation with my professorial friend moves into a discussion about university administrators, I cannot help noticing that the person on my right is sitting closer to me than is necessary. I hear a soft voice whispering, "You're not planning on sleeping alone tonight, are you?" Not knowing local custom but having experienced the persistence of Nairobi sales pitches, I slide to the left and emphatically assure the lady that I am going to be spending the night alone.

I resume the conversation with the professor, trying hard to ignore the person next to me. About 10 minutes later the whisper begins again: "You know, Ugandan women are different from Kenyan women." My curiosity gets the better of me, and I ask, "How?" "Kenyan girls are circumcised. They feel nothing." I must confess to being somewhat flabbergasted by this turn of conversation, and my reply lacks depth. "Really?" I say. Her rejoinder is to reel off the names of five or six Kenyan tribes, including the Kikuyu and Masai, who practice clitoridectomy, or female circumcision. I offer a few words of sympathy for the girls who feel "nothing." Victoria is unsympathetic, and I rejoin the general conversation, which has turned to comparing local beers with European beers.

About 10 minutes later the voice on my right begins again: "Are you sure you want to sleep by yourself tonight?" I excuse myself, go outside, find a taxi, and get back to my hotel, alone. It is still early in the evening and a good time to continue reading *Facing Mount Kenya* by Jomo Kenyatta, who is mostly remembered as the first president of the country and the person responsible for the emergence of Kenyan society in its contemporary form. Few are aware of his earlier career as an anthropologist and student of one of the founders of anthropology, Bronislaw Malinowski. Kenyatta studied with him at the London School of Economics during the 1930s.

The book I have before me was published in London in 1938. It is an extraordinary work from a social-science point of view. To my knowledge it is the first anthropological work on a tribal culture written by someone born and raised within that culture. As a trained anthropologist, Kenyatta was able to turn his intellectual weapons on the Europeans who had seized his beloved Africa.

> But a culture has no meaning apart from the social organization of life on which it is built. When the European comes to the Gikuyu country and robs the people of their land, he is taking away not only their livelihood, but the material symbol that holds family and tribe together. In doing this he gives one blow which cuts away the foundations from the whole Gikuyu life, social, moral, and economic.... Along with his land they rob him of his government, condemn his religious ideas, and ignore his fundamental conception of justice and morals, all in the name of civilization and progress.

After the conversation earlier in the evening I come with great interest to Chapter 6, which begins: "The custom of clitoridectomy of girls, which we are going to describe here, has been strongly attacked by a number of influential European agencies—

missionary, sentimental Pro-African, Government, educational and medical authorities." In this chapter the author undertakes both to describe in detail and to justify the procedures undergone in the initiation rites of boys and girls as practiced in the 1930s. Since male circumcision was widely known and practiced in Europe, Kenyatta focused on the female rite. "The real anthropological study, therefore, is to show that clitoridectomy, like Jewish circumcision, is a mere bodily mutilation which, however, is regarded as the *conditio sine qua non* of the whole teaching of tribal law, religion, and morality."

The ceremony of *irua* (circumcision) for both sexes took place in an age group, and members of this cohort maintained a special relation to each other throughout their lives. *Irua* was the rite of passage that admitted children into participation in the adult life of the community. For girls, circumcision was carried out before menstruation began. For 10 days before the operation there were rituals and special foods for the initiates.

On the day of the operation the girls received a ritual meal and spent half an hour bathing in cold river water to numb their feelings. A horn was sounded, and the girls came to a designated site, where a cowhide rug had been spread on the ground. No men were allowed anywhere in the vicinity. Each girl sat down with spread legs, and her sponsor, an older woman, sat behind her, legs entwined with hers to hold her for the operation. The coldest water available was used to douse the external genitalia. The girls looked skyward and did not blink. Then the *moruithia*, the ritual circumciser, came out of the crowd of women. Her face was painted with white and black ocher, and she had rattles tied to her legs. She took a special razor from her pocket and cut off the tip of each girl's clitoris. Each operation was performed in a single stroke. The wounds were treated with milk and herbs, and the girls were taken to a special hut for a period of healing.

Kenyatta stresses the social, ethical, and cultural aspects of *irua* as an integral part of an entire life pattern that was Kikuyu culture. I don't have any statistical information about current practices, but if my Ugandan informant is to be believed, it is still widespread among some groups in Kenyan society.

The next morning, having finished Kenyatta's book, I am off to the Nairobi Library for further information. Remarkably little is available there. The best source is *Obstetrics and Gynaecology in the Tropics and Developing Countries* by J. B. Lawson and P. B. Stewart (Edward Arnold Publishers, London, 1974). In a chapter entitled "Gynatresia," they report on widespread female circumcision in East Africa, sometimes involving excision of the labia minora as well as the clitoris. The practice has been outlawed in Egypt and the Sudan. The remaining question of how widespread the procedure is in Kenya is unresolved. I could go back to Buffalo Bill's to conduct a survey, but I have a mountain to climb and must leave this research to others.

28

Mammals I Have Known

I HAVE A NEWFOUND RESPECT for the hippopotamus. Having previously seen animals of this species only in zoos, I thought of them as ungainly, slovenly overeaters waddling slowly through life and surviving because they were too large and thick-skinned for any predator. But walking along the Mara and Tsavo rivers in Kenya and watching these creatures in their native habitats gives one a different perspective. They spend their days mostly submerged to help regulate temperature and their nights grazing in the woods and grasslands near the river. The sound of hippos snorting within earshot of one's tent at night is, I assure you, a genuine communion with nature.

Our guide several times warned us that hippos are dangerous and to be avoided on land; indeed, we were accompanied by armed rangers. He also told us that they occasionally come charging up the embankment when they are disturbed. One of the hikers inquired how fast the animals can move on land. The perceptive if non-numerical answer was "Faster than you."

As we rounded a bend in the river, the ranger sensed some unease on the part of the hippos bathing there and motioned us to move farther away from the water. Suddenly two of the animals started charging for the shore, and we quickly climbed a steep, rocky neighboring hillside. They stopped their charge at the embankment by the water's edge. I have not only a newfound respect for the hippopotamus but also an agility at rock climbing that I never knew I had.

Hippopotamus amphibius is a large mammal, weighing in at as much as five tons. The Linnaean name *hippopotamus*, derived from the Greek, means "river horse." Hippos have enormous heads with slitlike nostrils that can be closed underwater, short ears that twitch on surfacing, and prominent periscopelike eyes similar to those of a frog, so that when they are threatened, little more than their eyes and nose need be exposed above water level. They can remain submerged for as long as five minutes.

After observing hippos from a number of less exciting viewing points, I have had a chance to think about some aspects of their behavior. When they surface, they spout in a manner reminiscent of humpback whales. On reflection, one might expect some similarities between whales and hippopotamuses, as they represent alternate paths in evolution from land mammals to water dwellers. In the case of whales, there is some uncertainty about their ancestry, but the line of descent of the hippopotamus is better understood. Among the mammals, those of the order Artiodactyla—even-toed ungulates—are among the most recent and the most successful. The suborder Suiformes, consisting of non-ruminants, includes three families: pigs, peccaries, and hippopotamuses. In short, hippos are pigs on their way to becoming whales.

Curiously, in at least four mammalian orders there has been a return to marine existence. The long evolutionary path from fish to lungfish to amphibians to reptiles to land-dwelling mammals has, with respect to habitat, been reversed, giving rise to marine or amphibious mammals in four independent evolutionary lines. Such is the flexibility of the evolutionary process.

Among the carnivores are two closely related groups, the terrestrial carnivores and the aquatic Pinnipedia. The mammals of the latter group, which includes sea lions, walruses, and seals, breed on land but feed on fish, squid, shellfish, and occasionally penguins. They can be thought of as constituting an evolutionary line similar to what you might expect if dogs and cats had returned to an ocean habitat.

Another group among the mammalian orders, termed the primitive ungulates, includes aardvarks, elephants, and hydraxes. Also found in this group are the sea cows—manatees and dugongs—which are exclusively aquatic, feeding on vegetation in tropical rivers and coastal waters.

A great deal of physiologic and anatomic adaptation is required for a land mammal to take up life in the water. And yet in four independent cases, substantial evolutionary changes have given rise to a major order of amphibious and marine mammals. The capacity of living forms to evolve to fill every ecological niche is always impressive and nowhere more so than in East Africa, where the rich variety of mammals leaves one with a childlike sense of awe and wonder.

Of course, this is a part of the world where one naturally thinks a lot about evolution in any case. My campsite is only a few hundred miles from Koobi Fora, where some of the earliest hominid fossils were discovered. Some scientists believe that it was around here some 200,000 years ago that *Homo erectus* gave rise to *Homo sapiens*, forever altering our planet. Biochemical evidence based on DNA sequences is now being added to paleontological and archeological evidence, and one senses that within the next few years we shall arrive at a much firmer understanding of where we came from.

I write these words by lamplight in my tent near the Tsavo River. I'm traveling light, and my library consists of only one book: *Guide to Living Mammals* by J. E. Webb, J. A. Wallwork, and J. H. Elgood (Macmillan Press Ltd., 1977). The eastern sky is beginning to light up, and hearing neither the snorting of hippopotamuses nor the roaring of lions, I unzip the tent flap and peek out. Across the river a troop of baboons can be seen in the trees. As I come out of the tent, they begin to talk to me in a body language I do not understand. By watching these primates

I can sense our close relationship. I try to respond to their body language in kind, but no firm communication is established. As the sun comes up, the snorting of a hippopotamus upstream of the campsite can be heard. But it is far enough away; neither the baboons nor I am alarmed.

29

Daibutsu and Descartes

IT IS FOUR O'CLOCK in the morning, and I am lying in bed in the Hotel Continental in Yokohama. A number of thoughts are unfolding, doubtless related to jet lag, surviving the feared puffer fish at a delicious dinner last night, and visiting Daibutsu, a very large Buddha statue, at Kamakura.

Sometimes while lying awake in the wee hours of the morning, I think of René Descartes. In *Men of Mathematics,* by Eric Temple Bell, which I first read sometime in my youth, it is reported that Descartes produced his best original mathematical and philosophical reasoning while lying abed in the morning. When he later moved to Sweden to serve as advisor to its queen, he was forced at the royal whim to meet with his patron early every day, and his creativity waned.

However, thinking about Descartes in Yokohama has a different feel from thinking about this French philosopher in Fairfax, Va., or Woodbridge, Conn. I somehow cannot get Daibutsu out of my consciousness, nor can I fully separate thoughts of Descartes and Buddha. There is a nexus between the views of these two very different savants. Both resolved to probe the very depths of thought to understand the nature of existence, and both invested great intellectual energy in the task. René Descartes and Siddhārtha Gautama, the historical Buddha, were primarily concerned with getting rid of all uncertainty and penetrating to the core of reality.

For Descartes, the moment of truth came with the terse *Cogito, ergo sum* ("I think, therefore I am"). Note the radical nature of the "I," which occurs twice in the five-word English translation. The philosopher could deny anything about his thoughts, but the thinker or the mind of the thinker was to him an irreducible minimum. And so began the mind/body duality, which has haunted Western philosophy ever since Descartes's fourth Discourse became known.

Buddha had also spent a long time in deepest thought some 2,200 years before Descartes. After a struggle to find truth from several teachers, he rejected their methods, sat under a fig tree, and looked inward to seek truth. When he then attained "enlightenment," he proceeded to teach his method to the world. Part of it is a denial of the ego, which disappears into the five temporal and perishable factors of body, sensation, perception, mental phenomena, and consciousness. That doctrine has been part of Buddhism ever since. I recall that the scholar Hans W. Schumann wrote that for Buddhists it would be a fallacy to deduce the existence of a soul from the process of cognition. Rethinking this issue, I wonder if Schumann had Descartes in mind. In any case, Buddha sitting under the tree in broad daylight saw himself as part of the continuum and talked of the interrelatedness of all things.

Thus, the reflective thought that led Descartes to the undeniability of the thinking "I" led Siddhārtha to the denial of that "I." The Western mind/body paradox does not emerge from the Eastern interrelatedness of all things. Radically different views of the world are offered.

And so, cogitating in bed in the early morning hours, it seems

natural for me to inquire to what extent Descartes's thoughts
came from his work habits. Supine and thinking when the rest
of the world sleeps is a lonely occupation and may lead to solip-
sistic views of philosophy. "I think"—with a stress on the "I"—
seems to underplay that I think with language, which is humanity's common heritage, and that I think using methods
learned from parents, teachers, and others with whom I have
interacted. The "I," or thinking mind, is not as isolated from the
rest of the universe as it feels while lying in bed in the dark.
Descartes's necessity of having to "do it my way" underempha-
sizes the social character of knowledge and ignores that truth
itself is sometimes decided by a vote. When the contemporary
philosopher Thomas Kuhn refers to the acceptance of a paradigm,
he means that the majority of workers in some field of science
accept a view as the contingent truth.

Buddha, like Descartes, entered his meditations with a set of
preconceptions that did not disappear along the road to his enlightenment. He assumed as a given that the soul was permanent,
concluding that therefore the temporal activities identified with
the five categories listed above could not be the soul or the self.
For Siddhārtha, truth was eternal and therefore *not* accessible.
This rejection of essential parts of metaphysics leaves us no independent way of evaluating his message: We either take it or
leave it.

It is now five o'clock in the morning, and I see no light in the
east. I must confess a certain pleasure in the hubris of being able
to argue with Descartes and Siddhārtha, two of the all-time giants
of human thought. I believe this is possible because the past few
hundred years have seen the development of science, an enormously fruitful approach to all questions, including the eternal
queries. To carry out modern scientific reasoning, it has been
necessary to give up certain preconceptions from humanity's
past. We have had to surrender the priority of individual knowledge in return for social and cultural knowledge, and we have
had to give up permanent truth in return for transient, contingent
truth. As a result, we can avoid the paradoxes encountered in
the past. It is another case of standing on the shoulders of giants.

If we give up the *cogito* of Descartes as a starting point, we
need not focus on a disembodied mind and the paradoxes of mind/

body duality. If we set aside the necessity of permanence, we need not give up the ego that is both changing and continuous. We have come to know that all structures are a balance between growth and decay, and complexity itself emerges from these processes. This is not to say that all of the great philosophical questions have been answered. It says that perceptions have changed, and we are not tied to past categories to begin our inquiries.

Yesterday we got off the train at Hase and walked up to the shrine at Kamakura and stood before the great statue of the Buddha. I thought of this man whose words have inspired hundreds of millions of people and continue to do so. I find I relate easily to the great thinkers of the past. In the early morning hours I even argue with them. While standing in admiration, I had noted a small booth at the side of the statue. For another 20 yen one was permitted to go within.

The interior of Daibutsu was crowded with pilgrims who wanted to see the great Buddha from another perspective. I had never viewed the world from the entrails of a great savant, so it was a strange and awesome experience. I looked directly up into the head, and there was a large empty space. Ah, I thought, this is my koan, the Zen riddle I carry away with me from Kamakura. And so this morning I am working on my koan, and who would have thought that it would involve philosophers such as René Descartes and Thomas Kuhn? Well, in these rushed days, one takes enlightenment where one can get it.

The sky begins to light up ever so slowly, and I remember that I am in Yokohama to give a talk on research models. I begin to review my manuscript, but the distractions of Descartes and Daibutsu are too great. I get up, dress, and walk along the harbor watching the sunrise over an ocean that I usually note at sunset. I keep hearing the discordant voices of *Cogito, ergo sum* and *Om manipadme hum*.

30

Tatami Taxonomy

As I sit on a woven mat and savor the great variety of small dishes of food that are set before me, the slight aching of my legs is offset by the mild anesthetic effect of inhaling and sipping the tiny cups of hot sake. Ever the biologist, I begin to wonder how many major taxa of flora and fauna have gone into this meal. Japanese cuisine must be the richest in the world in terms of taxonomic variety, and the idea arises of exploring the world of living in terms of the menus of Nippon. This exercise in systematic gastronomy will be called, quite naturally, tatami taxonomy.

To make the exercise more tangible, I recall the norimaki sushi hors d'oeuvre that began this feast. This wonder of the life sciences contains materials from the four major classification groups, which perhaps had best be reviewed, just in case. The major division in all life forms is between the prokaryotes and

eukaryotes. The former include bacteria and simple nonnu-
cleated, unicellular forms. The latter encompass all higher forms
and are partitioned into four kingdoms: plants, animals, fungi,
and protists. Protists are a collection of algae and protozoa, which
are generally at a lower level of organization than plants and
animals.

Returning to the sushi, the major component is rice (*Oryza
sativa*), a higher plant. The rice surrounds a pink shrimp (*Pen-
aeus duorarum*), representing the animal kingdom. The entire
center is wrapped in sheets of nori seaweed (*Porphyra hitoe-
gusa*). Seaweeds are higher algae and thus fall within the protist
grouping. Finally, the prokaryotes have contributed the vinegar
that is mixed with the rice and has been fermented by *Acetobac-
ter aceti*. Clearly, we have a four-kingdom dish, and a very tasty
one at that.

Four kingdom foods are not uniquely Japanese; for example,
mushroom pizza fits the category. The plant kingdom contributes
wheat (*Triticum aestivum*); the cheese comes from the animal
cow (*Bos taurus*) and has been fermented by a prokaryote (*Strep-
tococcus lactis*). Fungi are seen in the common commercial
mushroom (*Agaricus bisporus*). See how easy it is to get into this
exercise?

Returning to Japanese cuisine, the true diversity is not in the
kingdoms but in the number of phyla, classes, and orders that
are aesthetically displayed and then eaten. To see this we must
look at the eukaryotic kingdoms separately, beginning with
protists.

The diverse protist grouping has as its major subdivisions algae
(excluding the blue-greens), slime molds, and protozoa. Algae, as
seaweed, appear frequently on the Japanese table, and perhaps
as many as 20 species are eaten. No other cuisine has exploited
this food from the ocean anywhere near as extensively as is done
in Japan. Among fungi, there is a rich variety of mushrooms as
well as of the related morels and truffles.

The higher plants comprise two major groups: bryophytes and
tracheophytes. Most edible species are tracheophytes, while bryophytes are mosses and liverworts. Major groups among the tracheophytes are the ferns, the gymnosperms, and the angiosperms. All
the usual commercial fruits and vegetables are angiosperms, and

these are represented in rich variety on the Japanese table. The other groups also contain edible forms, and salads may be graced by mountain ferns (*Osmunda japonica*). The gymnosperms include the ancient ginkgo tree along with conifers and cycads. Ginkgo nuts (*Ginkgo biloba*) are served in Japan.

The greatest variety in Japanese cuisine comes from the metazoans, also known as Animalia. This richness in species is the result of utilization of the marine forms that surround this island nation. We start with the major division between vertebrates and invertebrates. In dealing with vertebrates, the taxonomic system gets a bit complex, and I hope I will be forgiven for simplification. From among the enterocoelomates, the Japanese table is graced with sea urchins, sea urchin eggs, and sea cucumbers. The schizocoelomates include mollusks, arthropods, and a number of minor groups. Here we get into some of the major items of the sushi bar.

All of the three groups of mollusks provide seafood delicacies. Among the gastropods, or snails, the most popular here is abalone (genus Haliotis). Other snails and conchs are also used for food. The bivalves best known to chefs are clams, oysters, and scallops. The cephalopods on the menu include octopus, squid, and cuttlefish.

The arthropods are the largest taxonomic class and contain more species than the entire remainder of living forms. They are partitioned into four major groupings: arachnids, myriapods, insects, and crustaceans. Most of the edible animals are crustaceans, although a number of insects are eaten in various cultures around the world. Among the three groups of crustaceans, the major species of gastronomic interest are decapods: shrimps, crabs, lobsters, and crayfish.

Vertebrates are partitioned into fish, amphibians, reptiles, birds, and mammals. The fish divide into bony, cartilaginous, and jawless. Sharks are the best-known cartilaginous fish on the menu. Bony fish are in many ways the centerpieces of Japanese meals. They come in such a rich variety and are prepared here in so many cooked and uncooked states that a visitor should perhaps just best enjoy while trying to learn the fundamentals of ichthyological gastronomy.

It is obligatory at this point to make mention of the family

Tetradontidae, or fugu. These fish may contain the highly poisonous tetrodotoxin in certain tissue that must be removed by a licensed chef before the fish is served in restaurants. On a single try, I found it a rather tasty dish, but I would want to get some data and do some probability calculations before becoming a regular fugu eater.

Among birds and mammals, there are few surprises on the Japanese menu, so I turn my thoughts to other cuisines. The major mammalian classification is monotremes, marsupials, and placentals. With the first two my experience is limited. Once in Sydney, Australia, I saw kangaroo tail soup on the menu.

Among the orders of placental mammals, most have representatives that are used as food, either as game or as domesticated species. Artiodactyla, encompassing cows, pigs, and sheep, is the major order of agricultural importance. Perissodactyla, represented by the horse, finds its way into some national cuisines. I'll refrain, from this vantage point, from discoursing on the two orders of whales in the hope that this food source is being relinquished. Carnivora, as sea lions and walruses, occurs in the diets of far northern peoples. As for Rodentia, guinea pigs are served in the Andes, and I was once invited to have barbecued beaver in Michigan. Lagomorpha comes to us as hasenpfeffers.

Well, my tatami has become a magic carpet as the mind wanders around the world. There is a surprising overlap between what the taxonomist classifies and what the chef prepares. That's what it's like for us omnivores who eat high on the trophic pyramid.

This all raises some issues of great ethical and philosophical depth. I had best set down my chopsticks, sip a little more sake, and meditate a bit. I have, at the very least, thought of a new way to teach biological diversity.

31

Madame Pele

CALL ME A VULCANOPHILE. While Ishmael may feel the call of the sea, I am drawn to the sights and sounds and brimstone odor of molten lava pouring from the ground. Watching this birth of new land is for me a religious experience, the alpha and omega of planetary death and resurrection all combined in a single event.

It is difficult to arrange one's vacations around such episodic events as lava flows, so great flexibility is necessary. It is New Year's Day, and I find myself on board a plane flying from Ka-hului, Maui, to Hilo, Hawaii. I have no automobile or room reservations. In short, in a Thoreauvian sense, I am free. But to be free today is either to be rich enough to buy one's way out of any snags or to be willing to take a backpack and live in any setting that nature provides. I fit neither category, being a victim of a modicum of middle-class respectability, the great enemy of freedom. My only chance for liberty is an occasional attempt at Emersonian self-reliance.

The first challenge comes quickly. There are no cars for rent on the island for the next five days save for a single Suzuki off-road vehicle. I opt for this transportation at an obscenely high rate and then proceed to Hilo to find one of the least expensive hotels in town, the old reliable Hukilau.

At dawn I'm off on the road to Hawaii Volcanoes National Park headquarters to check out sites for viewing the new, large, still active flow. There are three possibilities: Chain of Craters Road from the park to where the flow crosses the road, Chain of Craters Road from Hilo to where the flow crosses the road, or overhead.

The next stop is Volcanoes Helicopter, where the first news is that all flights are booked for the next two days. My name is placed on various waiting lists, and I take off down Chain of Craters Road.

The 28-mile ride from park headquarters is spectacular even without a live volcanic flow. At the beginning of the drive through a Metrosideros fern forest, I begin to take notice of my "wheels." At first I had had unkind thoughts about it because it was costing so much money, and I had overlooked the fact that it is a spanking new, bright-red, sassy vehicle with the model name of Samurai. There is no other traffic on the nicely paved Chain of Craters Road, and as the Samurai picks up speed I begin to experience the positive pleasure of driving, a feeling I have not had in many years of traveling the car-choked highways of the northeastern United States. Rolling along through the scrub grassland down by the sea, a tall-in-the-saddle machismo overcomes me and my red Samurai, and a burst of discipline is required to contain the speed of rider and ridden.

On the highway is a roadblock for cars, and farther along, a second barrier has a warning for pedestrians. Just a few yards ahead, lava covers the highway and one can see the flow from somewhere up in Kilauea extending down into the sea. The ground is covered with fresh black pahoehoe lava and the remains of burned trees. Starting near the roadblock one can hike to the shore and see the shiny black sand produced when the molten basalt hits the water and vitrifies.

A mile or so back from the barrier is a national park visitors' center, the Wahaula Heiau, where an ancient Hawaiian temple once stood. The information center is presided over by Auntie Lei. A month ago her house was destroyed by lava. She speaks of her loss in a calm, resigned manner, and our talk turns to Madame Pele, the volcano goddess. She is not a malevolent goddess, for Auntie Lei reports that although Pele takes property, she does not take lives, and that is what is important.

Although we vulcanophiles take modern geology and vulcanology very seriously, we also find it difficult to ignore Madame Pele and other volcanic deities. I know these eruptions occur because the Pacific tectonic plate is rotating and this island is passing over a deep lithospheric hot spot, but somehow Madame Pele remains a vivid local personality.

I remember some years ago being at Omar the Tentmaker's in Honolulu to rent a backpack and sleeping bag for a hike up Mauna Loa. Omar and I had been discussing the trip, and as I was leaving he called me back. "One more thing," he said. "Madame Pele does not like it if you pee on hot lava." I have certainly never offended her that way, and the volcano goddess and I have been on good terms. More devout local followers of Madame Pele placate the goddess by throwing bottles of gin into the molten lava.

Auntie Lei and I have a good conversation. She shows me pictures of her house before, during, and after the lava flow. The "after" photo shows just black rock, for the flow has completely covered the burned remains. Auntie is a very stoic lady; she knows the island was built entirely of lava flows. It is not possible to love this island, which she does, and at the same time to hate the hot lava. At least Pele gives sufficient warning for the people to escape.

I drive back to the office of the helicopter service and find, to my delight, that there has been a cancellation on the three o'clock flight. The copter view of the volcano begins at the shoreline, the area I have just seen from the ground. The aerial view affords some sense of the substantial area covered by the latest flow. Flying back toward the vent, we can see areas where red molten lava still glows through. The vent itself is in a large caldera of lava mostly covered with a thin gray film of solidified rock, with large molten bubbles occasionally breaking through. Under-

ground tubes carry the lava down the mountainside. The experts
have no idea how long the flow will continue.

The short helicopter ride is over, and I remount Samurai to
return to Hilo.

Early the next morning, it's off to Kalapana, where several
houses have been destroyed by Pele. There are remains of metal
roofs, and a truck and tractor are stalled axle deep in newly
hardened rock. Other artifacts are firmly embedded in this black
basalt. Near the lava line stands a house entirely unscathed, only
200 feet from one that has been totally demolished.

Most geological events occur so slowly it takes a lifetime to
notice any change; not so on the slopes of Kilauea. There, at
times, one can watch the advancing river of lava continually
changing the landscape. It is perhaps that lesson in the imper-
manence of all things that makes vulcanology so fascinating.
While we are now all aware of the great tectonic processes by
which the earth recycles its chemical stuff, they seem to be intellectual abstractions. To stand on hard rocks that were liquid
only a few days before engenders a very tangible feel for the ever-
changing nature of our planet. Then again, I may just love this
place because the always unpredictable Madame Pele is such a
vivid symbol of freedom.

32

The Sociology of Palo Alto Coffee Shops

MY PARTICIPATION IN A PROJECT at the NASA Ames Research Center has resulted in a few weeks' sojourn in Palo Alto, California. Like a proper Bay Area resident, I am out jogging and walking early each morning, and I finish my constitutional with coffee and a pastry or roll. The "track" is on University and Hamilton Avenues, a route that offers me a bewildering variety of coffee shops, bakeries, and restaurants. I have sampled this large selection and find, as no great surprise, that no experience in life is without deeper meaning; the coffee shops have much to tell us about class and caste in this affluent university community. This introductory study deals with only six of the caffeine vendors whose establishments dot the map of downtown Palo Alto.

Join me, then, for an early morning stroll east from Emerson Street on University Avenue. The first place we stop at

is Burger King. The coffee is 50 cents and is served by giving
the customer a paper cup and pointing to the dispensing machines: regular and decaf. The customers' dress code varies from
informal to thrift shop. There are fathers with children, mothers
with children, and a higher fraction of blacks and Chicanos than
in any of the other bistros we shall visit. There is also the highest
percentage of smokers; some of the other institutions do not
permit smoking. A few college students are in evidence, and street
people tend to dine here when they have raised enough money.
Occasionally, someone in jacket and tie comes in, but this attire
is clearly noticeable by its difference. Danish pastry, bagels, and
croissants are available with the coffee. The food is good value
without elegance.

One block up the avenue we come to Jim's Coffee Shop. Jim's
opens at 6:30 in the morning, a half hour before the others, and
is clearly home to the blue-collar class. Seen in the window is an
array of donuts, including my favorite, old-fashioned frosted. The
fare is solid: donuts, bacon and eggs, and pancakes. The patrons
include office workers, lower-level managerial types (ties but no
jackets), gray-haired women, truck drivers, repairmen, and the
like. The clothing is undistinguished, and the coffee is 65 cents.

When I was a lad in the East, establishments like this were
always run by proprietors named Pappas, but this is the West,
and the man behind the cash register is a cheery Oriental, who,
I assume, is named Jim. At the back of the restaurant is a large
counter separating the eating area from the kitchen. Looking
through, one can see the chef frying eggs. He could be Pappas.

Another block of ambling along the street and we come to
Suzanne's Muffins, where there is often a line waiting for take-
out orders. Suzanne's is a cultural olio featuring a wide variety
of freshly baked muffins: bran, corn, blueberry, and more exotic
types. The cultural breadth comes from catering to those who
on occasion wet their muffins with espresso or cappuccino. Suzanne's is mostly home to those who work in offices and stores.
The women wear flowered dresses, and the men, who are in the
minority, wear neckties. The coffee (regular) costs 70 cents. This
establishment also serves a second constituency of less formal,
more university-oriented types who have an enthusiasm for
muffins.

A few doors down is Il Fornaio Bakery. The customers here are different from those at Suzanne's. There are no flowered dresses, no ties, and no jackets. The patrons have come on bicycles or on foot, wearing expensive jogging and walking shoes. They may wear shorts, jogging attire, or other very informal but not cheap clothing. There are strong shades of university influence here, and traces of yuppie character can also be noted. The coffee is 75 cents, and the pastries and rolls all have rococo Italian names, such as *piadina di crusca*. At Il Fornaio almost everyone reads while sipping morning coffee. A few are perusing books, but mostly they are poring over *The San Francisco Chronicle*, all reading different sections, of course.

Another block down and we arrive at a recessed storefront with chairs and tables in the foyer. The window reads: "Boulangerie, Fresh Bread, Croissants, Sandwiches." Here the dress code seems strictest. The women are in flowered dresses or the uniforms of medical personnel; the men all have ties, and most wear jackets. The bill of fare covers a broad range, from very crusty French bread to very sticky coffee cake, all made on the premises. Bagels and muffins are also available. I sense that these customers come from banks, insurance companies, real estate agencies, and medical offices. The coffee is 75 cents.

The last of our sample is Caffè Verona, Restaurant and Bar, on Hamilton Avenue. Although not part of the University Avenue catalogue of shops, it is included here because the coffee is 95 cents and I want to cover the full range. If one arrives too early in the morning, a sign in the window says: "Chiuso—Closed." The patrons here have a definite Stanford University stamp. They are mostly informal in attire but occasionally quite chic. When fashion prevails, the dresses tend to be black and lacy, worn with three-dimensional stockings, but never flowered prints.

In contrast to the other coffee sellers, the breakfast food here is hidden from view, so you must be a member of the ingroup to know what to order. I never discovered how to join the in-group, but I suspect it is mostly through friends or by a long novitiate period in which one learns by making mistakes. In any case, the rituals of this establishment seem more complex than those of the others.

The patrons at the tables form several groups: those who are reading a somewhat broad array of books and papers, those who are writing on lined paper or jotting down comments on dot matrix printer hard copy, those who are dreamily looking out into space, those who are talking to each other in very animated fashion, and those who are introducing their young children to the appropriate behavior for this environment. When the stereo is on at Caffè Verona, the music is typically an opera by Verdi.

I must point out that this study covers only the hours of 6:30 to 8:30 in the morning; the character and social structure may change through the day. For example, at night Jim's is closed, Burger King is more student and middle-class oriented, and Caffè Verona is somewhat more formal and modified in other ways, because it is the only one of our establishments to serve wine and beer. The second caveat I wish to stress is how restricted these studies have been. This is amateur sociology, and the available time for research was very limited. Only one site could be tested per morning to avoid the first-cup, second-cup appraisal problem, and each had to be visited several times to obtain prejudices of some statistical significance. At least 20 neighboring coffeehouses have not been investigated at all.

Next we come to the crucial moment when the essayist must reveal himself and expose his personality by identifying a favorite coffee dispensary. Since the essay is an idiosyncratic literary genre, the reader must be given a chance to penetrate the psyche and soul of the author. I know my literary responsibilities, but I am stuck. I have dined with pleasure and a sense of belonging, more or less, in all of these establishments. If I were to settle in this locale, which I do not plan to do, I would not permanently adopt any one of these places. Rather, I would branch out into the many other nearby breakfast restaurants and begin a more serious study. It's hard to resist a challenge for scholarly work. Thus, although I am being somewhat tongue-in-cheek about my morning coffee, I do believe that these kinds of observations can be the grounds for significant sociology.

One result emerges with certainty: The ambience of each coffee shop is unique, and each caters to a noticeably different segment of the vast diversity of our officially classless society, even in Palo Alto.

VI

33

Dirty Neoplatonists

THE UNIVERSITY OF CHICAGO has always held a strange position in the pantheon of great American universities. Midwestern yet cosmopolitan, it has been the home of a highly innovative, if considerably idiosyncratic, collection of scholars in a variety of disciplines. Site of a great classical revival, it was described in the 1940s as a Baptist institution where atheist professors taught Roman Catholic philosophy to Jewish undergraduates.

I was therefore not surprised to hear, some years ago, the story of someone finding in a University of Chicago men's room the following graffito: "X is a dirty Neoplatonist." X was unknown to me, so I refrain from repeating the name of the target individual. At Harvard that handwriting on the wall would have been pretentious; at Yale it would have been theater of the absurd; but at Chicago it was serious business. In subsequent days I have pondered what is so bad about being a Neoplatonist. And at last I have finally been led to an answer by reading articles on artificial intelligence.

To set the tone, we need a very brief look at Western intellectual thought and the sense in which much of the intellectual history of the past 2,400 years depends on a difference of outlook between Plato and Aristotle. Plato, profoundly influenced by mathematics, thought of the world in terms of ideal and abstract perfect forms that lay behind the sloppiness of sensory experience. Aristotle the biologist, elbow deep in dissections, thought of the world in terms of experimental categories and inductive generalizations. At its extreme, Platonism finds reality by deriving the observed world from abstract ideas. Aristotelianism at its extreme finds reality by deriving the essences from the observed world. The two major philosophical approaches and variations thereof waxed and waned through ancient and medieval times.

Because there are many possible ways to derive statements about the world from abstract principles and there are no agreed-upon rules about formulating principles, the varieties of Neoplatonism are legion. Aristotelianism, being more closely tied to empirical approaches, does not admit to a host of "neo-Aristotelianisms," although that tradition is by no means without variations.

With the rise of modern science, there was a curious return to Plato. It has been rather well stated by Ernst Cassirer in *The Problem of Knowledge*. He wrote:

> In the early dawn of the Renaissance modern philosophy returned to that Platonic position. Here philosophical speculation and the first rudiments of exact science meet. To Galileo the "new science" of dynamics, which he founded, meant first of all the decisive confirmation of what Plato had sought and demanded in his theory of ideas. It showed that the whole of Being is pervaded through and through with mathematical law and thanks to that is really accessible to human knowledge. The first definite proof of that harmony of truth and reality upon which, in the end, all possibility of knowledge rests was afforded not by mathematics but by natural science.

The emerging science of Galileo and Newton was a curious blend of the two Greek views. For it began in observation of falling apples, swinging pendulums, and the like, passed to mathematical laws ($F = ma$, $F = G\, m_1\, m_2/r^2$, etc), or Platonic ideals, and then returned to contact with the sensory world, as in the trajectories of falling bodies. Positing equations from observations constitutes an Aristotelian search for essences. Going from the fundamental equations to predict the behavior of observed bodies is to move from Platonic ideals to sense data.

The model of Galileo and Newton became the standard of classical physics and was further perfected in the hands of Laplace and others. Confirmation of the predicted existence and location of the planet Neptune was a glorious latter-day triumph of the Platonic approach to celestial mechanics. The impact of Newton's original achievements on the world of humane letters was overwhelming and can be well sensed from Alexander Pope's words:

> Nature and Nature's laws lay hid in night:
> God said, Let Newton be! and all was light.

We restate for emphasis that the great success of Newtonian mechanics, with its Platonic prediction of orbits and trajectories from a very few fundamental laws, had an enormous effect on Western intellectual history. People in every branch of scholarly endeavor took classical physics as a model and became Neoplatonists, seeking in their own disciplines laws that would make those disciplines understandable.

The Neoplatonic approach, heavily laden with physics envy, became a favorite if unexpressed mode of thought throughout academia. The graffito writer had indeed sensed the broad sweep of Neoplatonism. But why the criticism?

The stark fact that keeps emerging with great clarity from post-Newtonian science is that physics is the only field in which the Platonic model holds, and it doesn't even work for all of physics. In other areas the evidence leads to laws and generalizations, but the great complexity of the consequences of the laws makes it impossible to be predictive in a Platonic sense. That's too bad, but it's the way the universe is structured.

Thus, there are simple systems, which can be completely described in terms of a small number of axioms that account for the observed behavior, and complex systems, which require at the minimum descriptions of a large number of local domain laws and a large library of facts. All systems, except for those covered by classical physics, are complex and require an Aristotelian approach to their understanding.

The new understanding of Plato and Aristotle also emerges from the contemporary view of artificial intelligence. Early in the computer age, the prospect emerged of teaching machines to function in a mode analogous to what we call thought in humans. This vague concept was explored by formulating rules of thought: formal logic, syntactic rules of language, and arithmetic. That Platonic approach to intelligence, while interesting in principle, never produced results of sufficient depth and sophistication to generate much interest outside the group actually working in the area. Again, it seemed like a failure of the Platonic approach, because the rules were uncertain and in any case they generated a range of possibilities vastly beyond our capacity to deal with them.

Starting with the Dendral program in organic chemistry, the artificial intelligence community began to change from rule-based systems to approaches that depended heavily on including vast quantities of experimental knowledge in the computer's memory. By shifting to this more Aristotelian outlook, expert systems have provided the impetus for the dynamism of modern approaches to artificial intelligence.

The fundamental inability to predict the behavior of complex systems from a few simple rules has not been widely appreciated, nor has its significance been grasped. There are within both the sciences and the humanities a substantial number of scholars who still work under the myth of Platonic completeness. That outlook hinders the broader approaches to those fields and creates theory in constant danger of drifting away from the primary data of the everyday world. The gap between the beauty of concepts and the messiness of experience leads to confusion in many areas. That confusion is what must have motivated the graffito writer to speak of dirty Neoplatonists. I'm not sympathetic to graffiti as a mode of communication, but the thought expressed is understandable.

34

Quadrivial Pursuits

THERE HAS BEEN A LOT OF TALK around universities lately about core curricula, the canon, the culturally excluded, and a variety of related topics. I both welcome and sometimes tire of these discussions because real engagement doesn't quite seem to take place. In an effort to see the world from a different perspective, let's try to think about the evolution of universities and seek some insights from the past.

In the Middle Ages, curriculum at all levels was determined in large part by the range of intellectual interest and the dominance of the Church in European education. As the universities emerged, a system of instruction was developed along the following lines:

1. Propaedeutics—Latin and Holy Writ. This was canon in the

strictest sense, establishing the ecclesiastical foundations of education.

2. The trivium—the lower level of the seven liberal arts, consisting of grammar, rhetoric, and logic. These were, and are, tools of thought to proceed to the next level.

3. The quadrivium—the upper-level liberal arts, consisting of geometry, arithmetic, music, and astronomy.

4. Philosophy—the highest level of education, which was much less narrowly spelled out than the other levels.

This curriculum remained remarkably stable through the Renaissance (about 1350 to 1600), although after the Reformation (which essentially began with Luther in 1517) the notion of canon underwent considerable change of character. Canon in its original sense meant a rule laid down by an ecclesiastical council. As the word has evolved, that sense has considerably weakened, but the implied authority has not been lost. "Canon" is not a democratic but a hierarchical word, and a more appropriate term should perhaps be sought. If we mean a shared cultural heritage, then we require some means of establishing that shared system. Given the enormous cultural diversity of the present-day United States, this is a challenge of major proportions, and I have not sensed that it is one that most academics are eager to face.

With the post-Renaissance world, which I date from the burning of Giordano Bruno in 1600, there has been an enormous growth of knowledge. The earlier curriculum, nevertheless, had great staying power, for it provided an intellectual core permitting exchange among Western Europeans and Americans with some higher education.

By the 1800s, the growth of knowledge began to so overwhelm the classical structures that shared knowledge gave way to specialization. After about 150 years into this era, knowledge has become compartmentalized and departmentalized, until we are at a point that Ortega y Gasset has characterized as "the barbarism of specialization."

All of this institutional fragmentation has come upon us as we now sense a deep need for intellectual integration to restore a kind of wholeness to our thought. And so we nostalgically search for core programs and canons to establish a basis for breaking down the barriers; however, we cannot agree upon one, and we lack the authority to establish the other. The last two noble attempts at the canonical were the Harvard Classics and the University of Chicago Great Books. I am a fan of both of these series, but surely no canon for modern America can be without the Upanishads, the Talmud, and the Koran—which both series lacked. The last discussion of religion I participated in, which took place in my laboratory, involved two undergraduates, a Buddhist and a Muslim. We truly live in the postcanonical age.

Having stated the problem, I am, of course, under some obligation to suggest remedies, and I would like to suggest three: one intellectual, one institutional, and one slightly whimsical.

First, I want to suggest that the knowledge explosion, although real, is manageable. The reason is that computers are an extension of our intellects, with unlimited potential to store information and undreamed-of possibilities to process and reduce that information to forms that we can wrestle with, from a humanistic, moral, and liberal arts viewpoint. This kind of integration has been dealt with in biology under the rubric of the Matrix of Biological Knowledge (see H. J. Morowitz, "Models, Theory, and the Matrix of Biological Knowledge," *Bioscience* 39:177, 1989), but I believe it can be extended from the natural to the social sciences, and from them to the humanities. Computers represent the possibility of a revolution of perspective as profound as the invention of writing or of movable type. This is clearly the long-term way out of the barbarism of specialization into an age of the integration of knowledge.

To make this idea more tangible, think for a moment of a historian sitting before a work station on which he can (1) call up any document available in any world library, (2) have access to machine translation, (3) project any stored photograph, (4) retrieve related materials by any of a large number of sophisticated indexing programs, and (5) be in direct contact with any other historian using a similar facility. This description is limited only by imagination, for software capabilities in some areas even today go beyond what is recounted. How vastly more integrative our scholarly pursuits could be in this new environment. Given computers and information technology, the knowledge explosion is a great challenge and promise, not a danger or threat.

Turning to the institutional aspect of integration, we have an immediate problem. Most universities are now organized along departmental lines that tended to set about a hundred years ago.

Therefore, promotions, tenure, and similar operational features are structured within some of the narrowest features of specialization. We lack ways to deal with scholars and teachers who wish to cut across departmental lines, methodological lines, and pedagogical lines. The institutions are thus constructed in a manner to discourage integrative thought, methodological innovation, and new ways of teaching and learning.

Assessment is a particularly knotty problem. As specialists, we know how to evaluate other specialists, but we are very insecure in approaching work that goes beyond our own limited grasp. This is understandable—indeed, it incorporates some reasonable humility—but it does leave us tied to the barbarism of specialization. The only way out of this bind would appear to be engaged dialogue among specialists to try to seek common features and ideas that go beyond the departmental and into the universal. I don't think that this is easy. The cost in time and institutional resources will be considerable. However, unless we seriously participate in faculty teaching faculty, how can we hope to achieve new perspectives?

In the interim, we need institution-wide assessment procedures for promotion and tenure that, from the beginning, go beyond the departmental and focus on broader scholarship. We not only should tolerate integrative thought, we should encourage it.

What concerns some is the problem of maintaining standards in integrative disciplines. They worry that being a generalist may be a refuge for a failed or incompetent specialist. This is a real concern, and we are going to have to find a solution.

Finally, some whimsical thought about whether an educational core is possible. If we return to the curriculum that has had the most lasting value, the trivium and quadrivium, we note that it can be envisioned as containing the least gender- and culture-burdened material in the curriculum. It deals with thought in the broadest of terms and could provide common referents for the study of other subjects. In contrast to present cores, which are departmental synopses, this program would cut across most disciplines.

Present attempts at cores and cultural literacies have been thought by some to have too close an association to trivial pursuits. Perhaps we could elevate quadrivial pursuit to a new pursuit, incorporating the tools of thought in the best sense. The trivium and quadrivium were without doubt integrative and universal. That may come as a shock to the most committed departmentalists.

Our attempts to think about education reach deeply into the past and make an effort to see into the future. There are strongly integrative pulls at each pole. It is the role of the university to be able thus to stand between what was and what will be. To do this, we have to proceed with the work of the present, and there is a lot to be done.

35

Twenty Books Clad in Black or Red

THE OTHER DAY I was planning on reading for relaxation and pulled out my old copy of Aristotle's *The Parts of Animals*. Every time I start to read this extraordinary man's work, the following lines from Chaucer's *Canterbury Tales* seem to come to mind:

> A Clerk ther was of Oxenford also, ...
> For hym was levere have at his beddes heed
> Twenty bookes, clad in blak or reed
> Of Aristotle and his philosophie,
> Than robes riche. ...
> But al be that he was a philosophre,
> Yet hadde he but litel gold in cofre.

Those words are a delightful description of the ideal scholar, whose most cherished possessions were bound volumes of philosophy. All of which gets one to thinking about Aristotle's position in intellectual history. For the past 400 years, starting about 200 years after Chaucer's death, Aristotle has, in my opinion, gotten a bum deal.

A variety of circumstances and a group of unsympathetic readers have combined to blur our understanding of the man and his works. I am absolutely amazed at how often at cocktail parties, banquets, and other social functions one hears some thoughtless, unfeeling, prejudicial remark about his savant, and one is forced to correct the misconceptions. There is a trade-off for setting the record straight. Some people can get quite touchy about having their sloppy philosophical and historical ideas amended.

A major difficulty arises from the fact that most people whose information comes from academic philosophy fail to appreciate that—among his many fields of expertise—first and foremost, Aristotle was a biologist. Of the extant writing of this Greek sage, about a third or more deals with the life sciences. Yet let's examine the canonical work of Aristotle's writings on most college campuses, *The Basic Works of Aristotle,* edited by Richard P. McKeon. This 1,487-page tome contains barely 55 pages of biological writing, a mere 10% or less of the famous Peripatetic's extant work in the life sciences. No explanations are found in the preface, where it is noted that "for the most part omissions are from the four biological works." What McKeon has done is the equivalent of publishing *The Basic Works of Einstein* and deciding that for the most part the omissions will be from his writings on physics. A rather distorted picture emerges.

G. B. Kerferd did no better by Aristotle in his article in *The Encyclopedia of Philosophy,* another canonical work. He lists the biological writings under "Works on natural history" and in a very long and detailed article has almost nothing else to say about the biology and philosophy of science that were central to the work of the Academy.

One problem arises from the fact that, of the three great conceptual approaches to science—observation, experimentation, and theory—experimentation was unknown to the classical Greek savants. They worked back and forth between observation and theory and therefore lacked the powerful weapon of falsification to prune wrong theories. This is doubtless a fault of the ancients, but we cannot blame them for not discovering simultaneously everything that we now know. The arrogance of the present consists in judging past ages in terms of what we now know. It is an approach that bars us from the understanding and appreciation of bygone cultures. This seems a high price to pay for a few cheap shots at ancient scholars.

Thus, for example, Aristotle is patently wrong when he sets forth:

> So with animals, some spring from parent animals according to this kind, whilst others grow spontaneously and not from kindred stock; and of these instances of spontaneous generation some come from putrefying earth or vegetable matter, as is the case with a number of insects, whilst others are spontaneously generated in the inside of animals out of the secretions of their several organs.

He is, however, interpreting observations, and it took another 2,000 years before Francesco Redi performed his "experiments," which showed that insects originated not spontaneously but from eggs laid by other insects on decaying meat. And it was only a little more than 100 years ago that Louis Pasteur finally put to rest all ideas of spontaneous generation. It is hard for us to imagine science without experimentation, but that concept had to be discovered by a long and tortuous path. Thus, the science of Aristotle is largely based on observation and on collections of data of varying degrees of sophistication and validity.

A second difficulty arose from St. Thomas Aquinas's work in so thoroughly integrating Aristotle's writings into church doctrine that they became identified with dogma rather than being examined as fallible, falsifiable science. Hence, some Renaissance thinkers rejected Aristotelianism as an indirect way of questioning the dogma of the church.

A third attack on Aristotelian thought came with the rise of the new physics associated with Galileo and Newton. The discovery of the usefulness of mathematics in physics brought Plato, with his ideal forms and love of mathematics, into focus. The dogmatic, theologically associated view of Aristotelian physics, with its many errors, was vigorously attacked by Galileo. By the time biology began fully to develop as a science, Aristotelian physics was in disgrace, and insufficient attention was paid to his biological writings. That attitude persisted until the first quarter of this century, when the biologist and polymath D'Arcy Wentworth Thompson translated some of Aristotle's biology into English and pointed out how important these works were in understanding the philosophical writings. These ideas were subsequently developed further by the philosopher of science Marjorie Grene.

This is not the place to discuss the 20 books, but for the sake of scientific literacy, I list Aristotle's major biological writings:

- *Plants*—a short essay on botany
- *The History of Animals*—a long data bank describing the animals known to the Athenian school
- *The Parts of Animals*—the beginnings of comparative anatomy and physiology
- *The Motion of Animals*—two short essays on animal motility
- *The Generation of Animals*—a long essay on embryology

   Four short works on:

- *The Length of Life*
- *Youth and Senescence*
- *Life and Death*
- *Respiration*

In addition, many of the psychological writings would today be regarded as physiological psychology.

Why, you may ask, do academics keep pulling out 2,300-year-old books and waxing eloquent about them? Is it just a pedantic flourish to impress the uninitiated? I don't think so; the Clerk of Oxenford was not that type of fellow. In order to understand who we are and where we came from, we need to probe our roots. Understanding a past age is not easy, but it is a necessary part of dealing with the current state of mankind. It clearly makes no difference to Aristotle how we regard his contributions, but I argue that it makes a difference to you and me. The search for the good, the true, and the beautiful did not just begin; it's been going on for a long time. If all of this sounds like an advertisement for the Great Books concept, so be it. Throw away your television and buy Aristotle—you'll be much better off. While you're at it, try Chaucer!

36

The Litmus Test

IT'S DIFFICULT TO READ through a newspaper these days without encountering reference to a litmus test. Whether the article deals with appointing a Supreme Court justice or the director of the National Institutes of Health, the notion of such a test keeps working its way into the political rhetoric. To those of us who first encountered litmus in cherished childhood chemistry sets, the notion of a political litmus test impinges on the psyche in a strange way. Obviously, a more detailed understanding of litmus is called for. It begins, appropriately enough, with the 1911 edition of *Encyclopaedia Britannica:*

> LITMUS (apparently a corruption of *lacmus,* Dutch *lacmoes, lac,* lac, and *moes,* pulp, due to association with "lit," an obsolete word for dye, colour; the Ger. equivalent is *Lackmus,* Fr. *tournesol*), a col-

ouring matter which occurs in commerce in the form of small blue tablets, which, however, consist mostly, not of the pigment proper, but of calcium carbonate and sulphate and other matter devoid of tinctorial value. Litmus is extensively employed by chemists as an indicator for the detection of free acids and free alkalis. An aqueous infusion of litmus, when exactly neutralized by an acid, exhibits a violet colour, which by the least trace of free acid is changed to red, while free alkali turns it to blue. The reagent is generally used in the form of test paper—bibulous paper dyed red, purple or blue by the respective kind of infusion. Litmus is manufactured in Holland from the same kinds of lichens (species *Roccella* and *Lecanora*) as are used for the preparation of archil.

A little further philological investigation reveals that one can trace the word back to the Middle Dutch *leecmos* or *lijcmoes*, although some suggestion has been made that it derives from the Old Norse *litmosi*, meaning certain herbs used in dyeing.

The *Oxford English Dictionary* finds the first English reference to *lyztmose* in *Arnold's Chronicle* of 1502. Apparently, the word referred to the dry lichen also known as *white Corke*. Scientific usage was apparently well established in 1803, when Sir Humphry Davy, writing in volume XCIII of the *Philosophical Transactions,* noted, "A fluid came over, which reddened litmus paper." By 1827, extensive use of litmus paper was noted in the writings of Michael Faraday: "Two of them surpass the rest, these are litmus and turmeric papers."

In modern usage, litmus refers to a family of compounds, extracted from the lichens Roccella and Lecanora, that contain azolitmin and erythrolitmin. The neutral material is pale violet, while at the acid pH of 4.5 it is bright red, and at the alkaline pH of 8.3 it is bright blue. Lichens are a symbiotic association of cyanobacteria and fungi in which the former supply energy through photosynthesis and the latter provide other environmental interactions to allow this association of organisms to live in rather extreme environments. I have not been able to establish the role that the litmus group of compounds plays in the life history of lichens.

Aqueous solutions contain, in addition to water molecules, small amounts of hydrogen ions and hydroxyl ions. In neutral solutions, the positively charged hydrogen ions and negatively charged hydroxyl ions are present in equal numbers, approximately two of each for every billion molecules of water. Solutions that contain an excess of hydrogen ions are acids, and those containing an excess of hydroxyl ions are alkalis, or bases.

Litmus paper, which is sensitive to small shifts in the acidity of solutions, was extremely valuable in the early days of modern chemistry, when experimenters tended to work almost exclusively with strong acids and strong bases. Around the beginning of the twentieth century, our view of acids and bases changed.

Although vinegar and hydrochloric acid are both acids, it is clear to anyone who has ever spilled the first in the kitchen and the second in the laboratory that they have very different properties. Similarly, baking soda and lye are both alkalis but are very different in their reactions. There are strong and weak acids and strong and weak bases. Chemists needed a way to quantify this feature, and in 1923 Thomas Lowry and Johannes Brönsted introduced the pH scale, in which the measure is the negative logarithm to the base 10 of the hydrogen ion concentration (hydrogen ion activity, for the more chemically precise). Acid solutions have pH values below 7, alkaline solutions have values above 7.

With increased understanding of acidity and alkalinity, litmus paper was replaced by a group of papers that measured pH by an elaborate series of color changes, and ultimately by pH meters that could measure acidity or alkalinity to three or four significant figures. The litmus test is no longer of great use to chemists, because science is now far more sophisticated and recognizes a continuous gradation of values, rather than a dichotomy between acid and base.

Oddly enough, while the actual litmus test was falling out of use in chemistry, the concept of a social litmus test was entering common parlance. The first such usage reported by the *Oxford English Dictionary* was in 1957: "Their possession is as good a cultural litmus test as any I can think of." This usage has expanded, so that there are now a number of issues subject to litmus tests for public officials, who are either for or against some policy, either red or blue; no gradations are allowed.

Thus, while scientific thought has moved to subtle gradations of acidity and most other measurables, social thought has moved to all or none, black or white. This is exactly the reverse of the "two cultures" view, in which science is perceived as being rigid and precise, whereas social and humanistic thought supposedly allows all shades and hues encompassed in the enormous complexity of human emotions. We are being treated to the opposite of C.P. Snow's view. In a litmus-test world, it is not true that "a rose by any other name would smell as sweet."

The notion of a nonchemical litmus test thus dehumanizes us, for it does not allow that range of opinions and values that must characterize all human interactions and situations. It is inadequate for gradations of social and political thought, just as the actual litmus test has been inadequate in chemistry for a long time. The question of what the litmus does for the lichens persists and is more to a scientist's liking.

37

Expertise

AFTER HAVING BEEN AN ACADEMIC for more years than I care to recount, I suppose I have come to have a love-hate relationship with the professoriat. A few things have emerged in recent months that have shaken me up.

It began in the Persian Gulf crisis, leading up to and culminating with Operation Desert Storm. During most of this period, there was little hard news and a great desire for more information on the part of many Americans. The radio and television networks responded with around-the-clock coverage (except on Saturday morning, when the usual inane cartoons took precedence). In the absence of facts to report and with long hours to fill, the tendency is to turn to expert opinion, and that is what the media did.

But who were the experts? There was really a shortage of

genuine specialists, and those with sufficient facts were often involved in classified projects. But even in that case, expertise is a fuzzy concept when factual information is in such short supply and good theory virtually nonexistent. The media turned to the universities, and I sensed that some stations were willing to accept as a Middle East expert a social science professor who had smoked a Camel cigarette, sung "The Sheik of Araby," or seen Bob Hope and Bing Crosby in *The Road to Morocco*. Similarly, a military expert was often an academic who had risen above the rank of sergeant or gone deer hunting with an automatic rifle. Fifteen minutes of fame was being proffered, and few could answer with the truth: "I really have nothing of value to say at this time."

An example will illustrate the point. About 50 hours into the war, I was listening to a radio station. The newscaster introduced a professor from an area university whose specialty was battlefield psychology. The announcer asked, "After the terrible beating that the Iraqi army has been taking, how will that influence their behavior?" The professor replied, "Well, that's a very mixed bag: Some will fight harder, and some will be more likely to surrender." That statement exemplifies the kind of expertise I am discussing—state the obvious and state it pompously and authoritatively. But isn't there something shameless about such behavior? Aren't we professors under obligation to have some content to what we say?

While listening to all this expert opinion, I thought of lines I had read in *Lafcadio's Adventures,* by André Gide:

> I keep on talking... talking! It can't be helped! Once a professor, always a professor—even when one's drunk; and it's a subject I have at heart. . . .

When all this occurred I was in an unsettled state of mind about academia resulting from my participation in a task force dealing in part with funding for the social sciences. A number of heads of professional societies gave presentations that reverberated with the "We don't get no respect" theme. This was followed by the plaintive "We don't get enough research funds." I mused on this and concluded that the respect that the natural sciences gets stems from (1) a set of paradigms agreed upon by the community of professionals, (2) a reasonable ability to make predictions in some domain, and (3) a perception of importance of the questions being asked.

Certainly, the social sciences easily pass the test on criterion number 3, but do quite poorly on 1 and 2. I had the occasion to ask the representative of a political science group, "Did any member of your society predict the 'decommunization' of Eastern Europe, which was clearly one of the most important events of the past half century?" The reply was in the negative. Thus, we have political history and political theory, but we do not have a fully developed political science.

I do not want to be too hard on my social science colleagues, who face enormously difficult problems. I just want to agree about predictive ability. These forecasts are used by governments in formulating policies, and we should have a good sense of what the uncertainties actually are. One cannot criticize scholars for disagreeing over the answers to such complex questions, but I sense that the cult of expertise robs scholars of any appropriate humility about the ontological status of their assertions. Pompousness preempts professorial perplexity.

The post-Persian Gulf doubts about the university community were reinforced by reading *Tenured Radicals: How Politics Has Corrupted Higher Education*, a troubling book by Roger Kimball (Harper & Row, New York, 1990). Whereas I had been focusing on the social sciences, Kimball attacks the humanities. He argues that the liberal arts, which once sought the good, the true, and the beautiful, were being subjugated by a view of the good that had emerged within certain social movements outside the humanities. Thus, literary criticism is being used not to analyze and appreciate literature, but to advance various overt and covert political agendas.

He further contends that various theories, such as deconstructionism, are devices to impede the quest for meaning and the pursuit of the good, the true, and the beautiful. Kimball asserts that this is done through a complex vocabulary of mellifluous verbosities, full of sound and fury, that signify very little. He gives example after example of the vacuity of arguments set forth in modern literary criticism.

The next reinforcement of my lingering doubts came from

reading Jonathan Yardley's review in the *Washington Post* of *The Idea Brokers: Think Tanks and the Rise of the New Political Elite* by James A. Smith (Free Press, New York, 1991). Yardley sums up Smith's argument:

> The result of all this has been a proliferation of expertise that contributes as much confusion as enlightenment to the public dialogue. Books and reports spew forth at a numbing rate; gurus-in-residence spout sound bites on the evening news; the think tanks compete for attention and influence with elaborate self-promotional campaigns.

I am beginning to suspect even worse: The experts are dealing in areas where we lack the techniques, even at a basic theory-of-knowledge level, of assessing validity. Professionals, in fact, do make predictions in domains where predictability has not been established as a valid concept. By conferring the mantle of expertise, we glorify opinion and reify it, on which we base subsequent action. Smith worries that this wall of expertise excludes the public and allows the introduction of political agendas under the guise of informed advice.

In both the social sciences and the humanities, from the right and from the left, political ideologies are being slipped in, disguised as the search for the good, the true, and the beautiful. All this is very disturbing. I truly love the idea of the university community of scholars and the accompanying search for meaning. It has been the substance of my professional life. I would call for a return to the good old days but for two reasons. Those days were exclusionary, elitist, and frankly not all that good. And I would sound as if I suffered from the old-codger syndrome.

Therefore, let us address the future, looking forward to the day when the professoriat in its search for truth will humbly avoid pomposity, eschew confusing jargon, and modestly indicate the uncertainties encountered in that noble search. Let us look to the day when the groves of academe will resound with open, reasoned dialogue over opposing viewpoints, even among students.

38

The Afternoon Session

LUNCH IS OVER. We find our way back to the lecture hall and, in a certain disorder, wend our way back to our seats. Left over from the morning are the notes, the material on the blackboard, the slides in the projector, and the stale air. It must be difficult to build a large room with no aerodynamic mobility so that all exchange is by diffusion only. However, the architect of this hall has succeeded remarkably well in designing a convection-free, crosscurrent-free structure that maintains a continuous thermal inversion. In constant collision with the ceiling and walls are all the molecules of carbon dioxide that have ever been exhaled in this room. I estimate that the oxygen concentration is roughly equivalent to that of Kathmandu.

As the last stragglers wander in, the chairman announces the first speaker. He is 6 feet 4 inches tall, and the microphone has last been set for someone who is 4 feet 6 inches. But the fact that we cannot hear the leadoff speaker is about to be obscured by the fact that we cannot understand him, except for the initial phrase he utters: "First slide, please." The lights are dimmed, and the shadow show begins. The slides have been carefully crafted of gray on gray, so that the eye is never offended by blacks or whites. Any conflicting values in the presented information are thus muted. In the eerie half-light cast by the projector through its gray filters, the long speaker and short microphone stand looking like an apparition from a story by Edgar Allan Poe. The slides change with great regularity, and my mind drifts.

I am reminded of the story of the famous electron microscopist lecturing at an international congress. True to his profession, the number of his slides was "legion" (see Mark 5:9), and the talk was a machine-gun-like rattle of "next slide, please," etc, etc. He finally looked to the projection booth and asked, "Is that my last slide?" A small voice answered, "No, sir, you are four slides into the next speaker."

On the afternoon under consideration, the slides are occasionally interrupted by viewgraphs, which, from anywhere in the room, appear as strange Rorschach blots annotated by graffiti. From time to time a shaky pencil is seen quivering in the projected illustration.

The audio system has an unusual property. It suppresses the speaker's voice while amplifying the rattling of his papers. I don't know how this is done, as it requires a Maxwell demon or some other apparent violation of physical law. While the explanation may be in doubt, the phenomenon is clear and present.

At its loudest, the speaker's voice is soft and soporific. I believe he is speaking English with an Eastern European accent, although for the first three minutes I cannot rule out German with a French accent, Italian with a Japanese accent, or Spanish with a Hebrew accent. There are suggestions that he speaks an invented language, but I'm sure he comes from somewhere.

Within minutes of the talk's beginning, the hypoxia, the dim gray light, the shadow show, and the soft monotonic voice speaking in tongues begin to have their effect. A perfect setting has been produced for mass hypnosis, and one by one the members of the audience begin to drop off. First those who score a 5 on

Dr. Spiegel's eye-roll test of hypnotic susceptibility nod and go under. Then the 4s, and the 3s can be seen biting their lips to stay awake as the 2s yawn and gasp for oxygen.

I am the chairman and cannot doze, for politeness and gentility require at least one question after each talk, and in circumstances such as these, the responsibility must inevitably fall on the chairman. In addition, the chairman must begin the applause at the end of the talk.

By now the 3s on the eye-roll test have all nodded off, and I am pinching my earlobe to produce as much wakening pain as I can stand. I feel I am fading fast, and a certain terror overcomes me. I look around the room. Thank goodness for the 1s; they are still busily taking notes. Who, I wonder, is the patron saint of the 1s?

The 2s are beginning to go under, one by one. I chuckle, and pick up the water glass in front of me and knock it against my front teeth. This proves to be a noisy and not very successful upper. There is within me a great regret over not having had another coffee at lunch.

Suddenly, I am launched into wakefulness by a new thought. How will I ask a question, since I don't understand a word he is saying? I am reminded of the time in the dim past when Alfred Kinsey first lectured on his sexual studies to an audience of biologists. As the question period began, some wag rose in the back of the room and inquired, "Does dinitrophenol uncouple the reaction?"

A stratagem appears, and I pull out the book of abstracts. In the dim light I find four key words in his contribution. From the tetrad of mystical totems, ATP, NADP, CoA, and Ca^{2+}, I formulate a question and freeze it in my consciousness, so that when the talk ends, there *will* be a question.

I feel myself nodding off, and it is indeed like being pulled down into a whirlpool. I struggle with every erg of psychic energy, but the pull of sleep is greater than the restoring force. I think "amphetamine" to try to force my psyche into a state of wakefulness.

The eyelids droop, the breathing slows. I put my index finger between my teeth to bite. But my jaw goes limp as I join the number 1s in a cacophony of snoring and audible breathing that is filling the room in a great symphony to Morpheus. Only the 0s are still awake.

And then a strange thing happens. I dream that I am chairing a session and everyone is asleep. I dream that in the final throes of my efforts to stay awake, I hear the speaker finish his presentation, and I begin to applaud vigorously. I am in fact clapping my hands. As the entire audience is released from its collective trance by the cacophony, there is a fine, healthy applause. The speaker, who has gone overtime anyway, thinks I have used a somewhat abrupt timing device. No matter; he is probably convinced that I am a boor.

I am barely able to rise to my feet and utter, "Thank you, Dr. Glyczstyklim, for a very informative presentation on this most interesting subject. Are there any questions?"

A profound silence joins the stale air filling the room. I then turn to the speaker and slowly intone, with rising inflection, "ATP, NADH, CoA, and Ca^{2+}." The scientist, very animated, launches into yet another five minutes of inspired glossolalia, and we pass on to the next speaker.

39

Literati

I HAVE FOR SOME YEARS NOW been an advocate of the optimistic view that lurking in our archives and libraries are insights, propositions, and generalizations that arc hidden in the sheer mass of material. In the biological domain, there has been a growing movement to get publications on-line electronically, in order to use computerized methods of retrieval and analysis to search for hidden meaning. Other fields of study are using computer tools in similar ways.

Sometimes the insights come more traditionally, when one is browsing and grazing in the library. I used to say "browsing" and leave it at that until I came across the ecological distinction that browsing refers to eating leaves and shoots—that is, above ground level—whereas grazing refers to eating grasses. Since the books I seek are, for some curious reason, all too often on the lowest library shelf, I have had to add grazing to browsing to describe library foraging. Thus, I am often found by students in the undignified position of sitting on the floor in the stacks, thumbing through books from the bottom shelf. In addition, random-access material comes in a great variety of ways—through the mail, hand carried, by fax, and by electronic mail.

In any case, the other day I found myself at my desk reading through a work called *Selective Bibliography of American Literature 1775–1900*. Written in 1932 by B. M. Fullerton, the book was "designed to serve as a guide to noteworthy achievements in American letters." With the curious penchant that I have for reading reference books, I was captured and spent the next few hours trying to get a sense of the 387 authors Fullerton had chosen as exemplars of American letters.

The humbling aspect of the experience was the realization that I had some familiarity with only 49 of the authors, a mere 12.7% of the collection. Included, of course, were old favorites such as Poe, Twain, Thoreau, Whitman, Alcott, Dana, Dickinson, and Hawthorne. I didn't immediately sense any obvious lacunae in this compendium.

As I started to read through the descriptions of the authors and their works, something about the collection began to seem strange. I suppose that having spent so much time around universities, I have come to accept the politically correct feminist view that women were excluded from the literary life of nineteenth-century America. Yet to my great surprise, Fullerton's book—written with an entirely different agenda—lists 93 women in the collection, or 24%. From Abigail Adams to Constance Fenimore Woolson, a quarter of the authors the editor chose to include were from the group that some of my colleagues tell me was almost excluded from the mainstream of American intellectual life. That makes one think about judging the past by present-day paradigms.

The question arises: Who are these female authors, and did they play a significant part in my own educational past? Let us start with Abigail Adams, wife of one president and mother of another. (Feminists would correctly point out that there have been no women presidents.) Her letters, published in 1840, are important documents in understanding the development of the

United States of America as an independent nation. I have enjoyed reading excerpts from them now and then.

Louisa May Alcott won fame and acceptance in her own lifetime for a series of novels published between 1868 and 1886. As a youngster, I certainly remember reading *Little Women*. It was one of those books around the house that my mother managed to fill the bookshelves with. One read them as a matter of course. Alcott was a name to be reckoned with.

Emily Dickinson was clearly a poet of first rank. No history or collection of American poetry can exclude the haunting verse of this recluse, who turned her thoughts inward to think of life and death—focusing, in my view, too much on the latter. It is in compendiums of American poetry that I most frequently encounter the work of this woman.

Another book that was part of my childhood was *Hans Brinker: or, The Silver Skates,* by Mary Elizabeth Mapes Dodge. This was another volume in that mini-library that surrounded me from the second to the 16th year of my life, and it had an obvious influence on my view of the world. Mary Dodge also left her mark as editor of *St. Nicholas Magazine,* which was an important voice in American literature in the nineteenth century.

Julia Ward Howe gave voice to the "Battle Hymn of the Republic," the emotional, moral statement about the Civil War from an abolitionist point of view. She wrote for the *New York Tribune* and the *Anti-Slavery Standard,* and edited the Boston *Commonwealth* and the *Woman's Journal.*

During her short life, Emma Lazarus won recognition for two extraordinary books of poems. Hers was the voice of immigrant America. She gained immortality through her words engraved on the Statue of Liberty: "Give me your tired, your poor,/Your huddled masses yearning to breathe free." Those lines are also engraved on the hearts of the descendants of those who found freedom on these shores.

Harriet Beecher Stowe was "the author of the most popular single novel ever written in America," *Uncle Tom's Cabin.* It sold over a million copies in 10 years, a remarkable achievement in a country of 26 million inhabitants. This was a novel that may well have influenced the course of history. Stowe was the author of a number of other novels and short stories.

Since I relate in a very personal way to these seven women authors, I find it difficult to maintain that they were not an important part of my literary heritage. I am mindful of the fact that only seven of the 49 writers I knew of were women, and this is clearly an underrepresentation. And though I thus accept that women had limited opportunities in a number of areas, I do not think that this is a useful way to look at literature. Fullerton has made it clear by example that in this case the politically correct point of view is overdone. Women were not excluded from the nineteenth-century world of letters in America. Interpretation cannot erase facts that were carefully gathered by a bibliophile as a guide to other collectors of the works of American authors. We must accept the data and then build our understanding. I would thus suggest that *Selective Bibliography of American Literature 1775–1900* should be thoroughly plumbed by feminist scholars for its view of women writers during the eighteenth and nineteenth centuries. I suspect that some of these authors have been forgotten, and this bibliophile's selection might lead to some interesting studies.

40

Potlatch

WHEN WE TAKE A CLOSE LOOK at the activities of men and women in "primitive" societies, that study is designated as anthropology, but we lack a similar professional label for the scientific observation of our friends, relatives, colleagues, and neighbors. There is something very boorish about staring at our peers and taking notes. Thus I once thought that a true cultural study of contemporary Terre Haute, Ind., Oxnard, Calif., or Fairfax, Va., didn't exist, although I knew Lewis Thomas had reported on the activities of the biologists on Stony Beach at Woods Hole, Mass. In general, there is a surprising reliance on questionnaires that ignore the primary existential fact that people lie, boast, and exaggerate. Consequently, I thought that there are features of society that are better understood in the Trobriand Islands than on Manhattan and Staten Islands. I have since become aware of modern studies by such scholars as John D. Dorst and Constance Perin, who focus on the anthropology of contemporary American life, and I must thus revise my views of cultural understanding. Such intellectual burdens are the price of discussing ideas with learned colleagues.

The previous paragraph serves as a prelude to a discussion of American mores and folklore conducted one night at the bar of the XYZ Yacht Club. I had not thought of the barstool as my chair in anthropology, but one must accept enlightenment wherever it comes. Of course at the time I had not read *The Headman and I: Ambiguity and Ambivalence in the Fieldworking Experience,* by Jean-Paul Dumont, so I was unaware that the anthropologist is never entirely removed from the events he observes. You might say I was engaged in "the persistent and tenacious myth of anthropological objectivity." But in any case, I was in an unobtrusive spot, sipping my brew and trying to record behavior without influencing that behavior by the presence of a known observer. Just as the great Bronislaw Malinowski became one of the leading anthropological theorists as a result of being stranded on a South Pacific island in World War I, so fate had brought me to this spot—indeed, to this Pacific island whose name I shall withhold in order, as you will see, to protect the guilty.

The central events began when someone bought drinks for everyone at the bar. Not wishing to stand out, I quietly accepted a second light beer and focused on the question of why individuals buy rounds for the house. Very often many of the recipients are strangers to the purchaser, and frequently the donor can ill afford his largess. I recalled once being in the pub of the University of Queensland in Australia. A ritual of everyone at a large round table buying a pitcher of beer for everyone else at the table began. The ceremony had an inexorable character about it and could lead only to the entire group's getting stoned. It all seemed so rule-driven. In addition, I was jet-lagged from an endless day— 18 hours—of flying west. "Ah sunflower weary of time, who countest the steps of the sun" (Blake).

As I sat pondering the rite of purchasing for the group, an acquaintance two stools down on the right announced that it was his birthday and bought a *very* expensive bottle of champagne for the folks sitting around him. When a glass of the bubbly was slipped in front of me, I realized that this kind of research was going to require more personal involvement and sacrifice than I had intended. I celebrated my being only walking distance away from my destination for the night.

As the bottle was emptied, my social science thoughts were replaced by more literary ones, and the lines of Omar came to mind: "I wonder what the vintners buy one half so precious as the stuff they sell." Given the cost of that particular bottle of champagne, I suppose that they buy diamonds, gold bullion, or Rolls Royces.

After the last drops dripped from the lip of the bottle, the birthday boy (as we used to say) inquired who was going to buy him another bottle as a present. The occupant two seats to my left, a perennial misanthrope who was in total and absolute ignorance of the cost of this rare French vintage, volunteered the honor. Another cork was popped, and a glass was placed in front of me. I was truly being asked to give my all for research and thought again of the hardships borne by poor Malinowski, abandoned on an island, not knowing how many years would pass before his return to his own civilization.

At this point, some hostility began between the man on my immediate left and the birthday celebrant. They had been acquaintances for 12 years and basically didn't like each other. Unpleasant words flew back and forth before my eyes and behind my head. All of this was interrupted when the man on my immediate right ordered another bottle of the same champagne. He had a large heart and small wallet but would not be dissuaded from his generosity. An unbelievable third glass of the precious brew was placed before me, and I had to worry about my objectivity, which is so important to scientific observers.

At this point, the grumpy purchaser of bottle number two found out what it was going to cost him and stormed away from the bar in a voluble rage. The two antagonists resumed their badmouthing in more and more strident tones. I sensed that my research was becoming perilous, so I paid for my one beer and left, after first whispering to my acquaintance two stools to the right, "Blessed are the peacemakers, for they shall inherit the earth." If I then felt bad about leaving my researches, I reason in retrospect that not even Dumont would have required such engagement on the part of the scientist. Mine was unheeded advice, for I later found that the two drinkers came to blows a few minutes after I left the club. At the very time when they were fist to fist, I was back in my room falling asleep with a copy of Ruth Benedict's *Patterns of Culture* clutched in my hand. I was hoping that this famous anthropological work would explain the evening's events, and further reading the next morning indeed helped.

One of the major sections of Benedict's book deals with the Kwakiutl Indians of Vancouver Island. She referred to their culture as Dionysian. One of the social inventions of these North American natives was the potlatch ceremony, in which wealth or the giving away of wealth was used to climb the ladder of titular names with their prerogatives. One climbed the ladder by giving away or destroying material possessions.

> The object of all Kwakiutl enterprise was to show oneself superior to one's rivals—uncensored glorification and ridicule of all comers. . . . The whole economic system of the Northwest Coast was bent to the service of this obsession. There were two means by which a chief could achieve the victory he sought. One was by shaming his rival by presenting him with more property than he could return with the required interest. The other was by destroying property.

Neither of these acts was of any real economic advantage, and they took place at potlatches. Benedict goes into great detail about the ceremony. What I had witnessed at the yacht club was a minipotlatch ritual. Individuals were wasting large amounts of money on champagne to establish their superiority. The process was interrupted by the fight. Of course, all of this is too simplistic, but knowing about the potlatch ceremony gives some insight into my study of barstool culture. I am encouraged to extend my studies to other aspects of our social behavior.

VII

41

What Is Life? Now Is the Time to Find Out

RECENT EXCITEMENT ABOUT sequencing the human genome has generated interesting possibilities for understanding general biology. The raison d'être of sequencing is to understand the biochemical behavior of an organism. The sequence of DNA base pairs must be indexed to the full structural and functional repertoire of the cells being studied. The time is now at hand to undertake a total analysis, not for something so complex as a human being, but for the smallest and simplest autonomous, self-replicating organisms. The hope is that basic principles will be most easily seen in the simplest system, in much the same way that early twentieth-century physicists chose hydrogen as the simplest system exhibiting the features of atomic spectra.

In 1962, there began an intensive study of the pleuropneumonia-like organisms, called PPLO, which were then (as now) believed to be the tiniest organisms capable of growth and replication outside a host cell. Almost three decades of further study have focused attention on one genus of the group, Mycoplasma, which includes cells as small as 0.3 μm in diameter and genomes that consist of loops of double-stranded DNA about 800,000 base pairs in length. Some of these cells contain only a billion atoms yet fulfill all our criteria for living organisms. They thus serve as ideal models for probing life at the most reduced level.

While the criterion of size is an obvious one, that of simplicity is somewhat more subtle and relates to the amount of genetic material. A double-stranded DNA molecule having 800,000 base pairs contains 266,666 nucleotide triplets. Since each triplet codes for one amino acid and there are, on average, about 600 amino acids per protein, we would expect such an organism to contain only 400 to 500 different kinds of protein molecules. Thus, we can postulate that all of the cell's activities have to be carried out with a manageable number of functions for computer modeling.

For a more concrete view, let us focus on one well-characterized species, *Mycoplasma capricolum*. This microorganism was originally isolated from goat lung and is associated with a pleuropneumonia. Cultures can be grown on a chemically defined medium containing 40 components. The cells are membranous vesicles containing genomes, ribosomes, enzymes, messenger and transfer RNAs, and various low-molecular-weight metabolites. The major cellular functions consist of extracting small molecules and ions from the medium, transforming the chemical potential of some nutrients to that of ATP, directing the synthesis of required small molecules, synthesizing the appropriate macromolecules, and, finally, growing and dividing to yield two cells for each original cell. These functions are close to the irreducible minimum necessary to qualify as a living organism.

The proteins of *M. capricolum* have been analyzed by two-dimensional electrophoresis, in which the first stage separates molecules by electric charge and the second by size. These gels show 405 different spots, 50 of which are due to ribosomal proteins and the rest to membrane protein and soluble proteins. The actual number of observed proteins is therefore within the range predicted from the size of the genome.

A crucial issue in understanding any cell is relating each portion of the DNA sequence to the proteins and functional RNA that it encodes. Then we must determine the function of each of these molecules and understand the integrated behavior of this entire collection of gene products. According to the reductionist view, this should form a complete understanding of the organism. *M. capricolum* seems to be an ideal system for this analysis, although other mycoplasmas are possible candidates.

The program can be carried out in the following way. First we must sequence the entire genome. With current techniques, an 800,000-base sequence is a very feasible project, and a number of laboratories are undertaking parts of it for various mycoplasmas. Each of the 405 protein spots can be subjected to preliminary analysis by sequencing a small number of amino acids from the amino terminal ends of the molecules. Each protein as well as ribosomal and transfer RNAs can then be located on the DNA molecules, yielding a complete genetic map in terms of the primary gene products. The remaining amino acid sequences of all the proteins can then be read out from the DNA sequence. The next task is to determine the function(s) of every gene product.

Determining protein functions can be carried out in the first instance by comparing amino acid sequences with all the various data bases of proteins and encoding DNAs. Guidance as to which enzymes to look for may be acquired by analyzing the input (components of the defined medium) and the output (known requirements for monomers, cofactors, and universally occurring small molecules) subject to intermediate metabolism. This information should provide a complete understanding of the organism's activities. Filling in all the missing data may require some additional studies of proteins and their functions using more traditional methods of biochemistry.

When all the enzymatic functions have been established, the various control loops and feedback systems must be determined by laboratory study as well as by comparison with similar systems in other organisms. After that task is completed, we should possess a flowchart of the organism in a chemical engineering sense. The chart must then be supplemented with kinetic parameters, such as binding constants and rate constants. When enough of that information is obtained, by a variety of experimental and theoretical methods, it will be possible to construct a computer model of the entire cell, so that every laboratory experiment can be paralleled by a computer experiment that should yield the same answer. At this stage we would understand *M. capricolum* as a chemical reaction system. This comprehension does not include or preclude the detailed molecular mechanics of the individual enzymatic or structure-forming steps: It represents them by reaction rates and binding constants. Within such limitations, the success of this program would constitute an understanding of life.

The question has been raised, Why mycoplasmas rather than other, much more extensively characterized prokaryotes? Mycoplasmas are genuinely simpler, in having a genome one-sixth the size of *E. coli*'s genome and one-half the size of the genome of the simplest competing bacterium. Mycoplasmas have no internal organelles and lack cell walls and periplasmic spaces. All proteins appear to be expressed in this species, and because of genetic simplicity the computer model will be far more amenable to theoretical analysis.

The proposal for fully understanding Mycoplasma is still small science by today's standards, perhaps costing a few million dollars over a period of five years. This project would provide a good foundation for the vastly larger schemes that are currently anticipated with such enthusiasm. The practical returns might be substantial. The opportunity to explore the nature of life fulfills a deeper need. In short, we are now able to do experiments asking the question, What is life? at a very deep chemical level. Let's move ahead and get on with it.

42

Brekekekex, Ko-ax, Ko-ax!

WHERE HAVE ALL THE FROGS GONE? That question introduces matters that are urgent, whimsical, and sad—and, as we shall see, that also have relevance to the interrelatedness of all things. My first contact with this query came in September 1989. I was attending a meeting of the Board of Biology of the National Research Council. During predinner socializing, Professor David Wake of the University of California, Berkeley, was telling us about the summer meeting of the International Herpetological Society in England. In the informal parts of that meeting, a number of investigators from around the world had made mention of declining frog populations in many locales. Wake was deeply concerned about the potential magnitude and significance of a loss that seemed to be affecting a large number of genera and species.

As the discussion unfolded, the possibility emerged that a significant ecological early-warning signal was being sounded. We became convinced that the anecdotal information needed to be strengthened, and the extent and nature of the problem evaluated. A prompt international meeting of experts on amphibians seemed to be the path of choice. By the time the dinner was over, such a meeting was planned, with Wake in charge of recruiting the experts and yours truly in charge of recruiting the funding.

On February 19 and 20, 1990, a workshop on "Declining Amphibian Populations—A Global Phenomenon?" was held at the Beckman Center in Irvine, Calif., under the sponsorship of the National Academy of Sciences. Thirty-five experts on various aspects of the problem were gathered from around the world to assess the global issue. In attendance were frog biologists, ecologists, embryologists, and physiologists, as well as a statistician, a frog pathologist, and an atmospheric chemist. If the meeting were being planned again, I would also recommend including an amphibian toxicologist.

From September to February, Professor Wake had communicated with these and other experts and established the Frog Log, an on-line collection of all relevant reports on frog populations. This activity provided an interesting example of scientists responding, in a rapid and totally informal manner, to a problem perceived to be serious and urgent.

As a consequence of my attempts at fund-raising among conservation groups, word of the coming meeting found its way into the *Washington Post*. The press picked up on the frog story, and by the time of the Irvine meeting there were 10 reporters on hand. Strong public interest in frogs is a subject we shall return to. I note for now that, overall, the press coverage was accurate and to the point. The one exception I encountered was a smart-ass, allegedly humorous piece by columnist Tony Kornheiser. The best news article I saw was by Karen R. Long in the *Cleveland Plain Dealer*.

The Irvine meeting was a model of scientific interchange. Its conclusions, reached by consensus, are both modest and significant:

1. Since 1978 there has been a significant decline in frog populations in many areas around the world. The greatest decline

has been in mountain species of western North America, and the least in low-altitude equatorial regions. The data are fragmentary and often qualitative and anecdotal; in the aggregate, however, they acquire significance. Declines and extinctions are species-specific as well as habitat-specific.

2. Most declines and extinctions can be attributed to local factors: acid rain, heavy metal mobilization, pesticide release, deforestation, hydrologic modifications, and introduction of exotic species. In short, most are due to anthropogenic factors, and there is no easy fix. The growth of human populations and the activities of humans appear to be responsible for most amphibian population decreases.

3. Amphibians are perhaps uniquely sensitive to environmental change. This is a result of their physiologic, behavioral, life historical, and morphogenetic patterns. Amphibians, which include frogs, salamanders, and the wormlike caecilians, are evolutionary intermediates between fish and reptiles. They thus tend to have both aquatic and terrestrial stages of the life cycle. Their skins are highly permeable. They tend to be at the third trophic level or higher (third-trophic-level organisms feed on animals that feed on primary plant material). All of these features tend to make amphibians especially sensitive to habitat changes.

Having analyzed the problem and arrived at an understanding, participants found the last few hours of the meeting rather frustrating. There was no national or international body to whom the results could be communicated. No one is charged with responsibility for the nation's or the planet's taxa. Since the factors in amphibian decline are, as noted, largely anthropogenic, the solutions lie in major changes of human lifestyles or demographics. Such changes go to the heart of the political process, to which the biologist has only a minor input. There is a proposal before Congress for a National Institute for the Environment, but that is obviously some years away.

I returned home and began to paraphrase John Donne: "No taxon is an island, entire of itself; every species is a piece of the ecosystem, a part of the main." With creative assistance from my amanuensis, Iris Knell, this ended: "And therefore never send to know for whom the frogs croak; they croak for thee."

During the conference I became impressed by the intense public interest in frogs. A public that had not responded very actively to the loss of biologic diversity per se was intensely interested when one very common group of organisms became the focus of ecological attention. After all, *Homo sapiens* and the many species of frogs have had a strange relation over the millennia:

> *Of all the funny things that live in woodland, marsh, or bog,*
> *That creep the ground, or fly the air, the funniest thing's a frog.*
> —Anonymous, "The Scientific Frog," circa 1865

One begins to wonder why the wicked witch turned the prince into a frog and why the princess was tempted to kiss the frog.

In all of this wondering, my thoughts turned to Kermit, and I was tempted to call this article "Where Has Kermit Gone?" However, before I had finished writing it, Jim Henson, the creator of the Muppets and the voice of Kermit, died suddenly, and the irony of his death at this time was too poignant for me to retain the title. I had met him only once, when he came to Pierson College at Yale and spent an evening visiting with students in the living room of the master's house. The world is greatly impoverished by Kermit's departure; the bell indeed tolls for us.

For my title I then turned to the chorus of frogs from the satirical drama *The Frogs* by Aristophanes. The relentless "bre-kekekex, ko-ax, ko-ax" is perhaps symbolic of our great difficulty in dealing with environmental problems sensibly. In attempting to trace our feelings about frogs, I also came across a line from Plutarch that provides some insight into the present situation:

> It was the saying of Bion that though the boys throw stones at frogs in sport, yet the frogs do not die in sport but in earnest.

43

A Rose by Any Other Name

IT IS NOT COMPLETELY TRANSPARENT what Gertrude Stein had in mind when, in the puzzling piece "Sacred Emily," she wrote: "Rose is a rose is a rose is a rose." She repeated the thought in *Lectures in America*, where she noted: "A rose is a rose is a rose." We might naively assume that Stein was lumping all roses in an act of taxonomic know-nothingness. More likely she was restating in her own syntax Ralph Waldo Emerson's thought:

> Man is timid and apologetic; he is no longer upright; he dares not say "I think," or "I am," but quotes some saint or sage. He is ashamed before the blade of grass or the blowing rose. These roses under my window make no reference to former roses or to better ones; they are for what they are; they exist for God today. There is no time to them. There is simply the rose; it is perfect in every moment of its existence.

I would not have pursued these flowery thoughts had I not heard Stanwyn Shetler of the Smithsonian Institution speak about the great difficulties of classifying roses, which became a matter of public note in the discussion of choosing the rose as the national flower of the United States. Through the kindness of Dr. Shetler, I obtained a transcript of the "House of Representatives Hearing on Designation of a National Floral Emblem." This matter of civic botany came under the purview of the Subcommittee on Census of Population of the Committee on Post Office and Civil Service. Congressman Robert Garcia of New York chaired the subcommittee.

Lest the reader think that this is a simple matter, note Garcia's introductory remark: "The designation of a national floral emblem lies at the center of one of the longest running debates [before Congress]." Indeed, the matter seems to have been considered by the federal legislature for at least 100 years. By June 25, 1986, the choice seemed to have been narrowed to either the rose or the marigold.

The Congressional Research Service of the Library of Congress had suggested the following criteria:

Essential

1. Native to North America.
2. Not modified through selection, breeding, or hybridization.
3. Easily recognized, attractive and showy.
4. Widespread—grows outdoors in every state.

Desirable

1. Easily propagated.
2. Widely grown in the world, to serve as an "ambassador" for the United States.
3. Possessing historical origins—prevalent and popular in colonial times.

So the matter became botanical and historical as well as political.

Senator J. Bennett Johnston of Louisiana led off the hearing. He eloquently offered historical background:

Christopher Columbus himself picked up a rosebush floating in the water the day before he discovered America. That, in itself, ought to be the omen of greatness of this country, established by Christopher Columbus himself.

When the Mayflower landed in America, it was recorded that "the shore was fragrant like the smell of a rose garden, and the happy children gathered strawberries and single wild roses."

Roses grew abundantly at Mount Vernon, and Martha Washington used them to make rosewater. The rose played a role in choosing the body to be placed in the Tomb of the Unknown Soldier following World War I, and Congress wore white roses to mourn the death of Franklin Roosevelt.

As the hearings wore on, George M. White, Architect of the Capitol, and Dean Norton, Horticulturist of Mount Vernon, made less than earthshaking statements. White noted, "There is no obstacle to the incorporation of whatever emblem should be selected as time goes on." Norton affirmed, "I think either the marigold or rose would be a wonderful selection."

With Shetler's testimony, the pro forma matters ended, and the botanical issues were addressed. First, "rose" is not an unambiguous designation. "There are 200 species of roses in the United States alone." I wonder if Gertrude Stein was aware of that fact. Not only are there over 200 species of roses, but their taxonomy is in a state of disarray. The reason, of course, is that the selfsame Congress that is concerned with a national floral emblem has never seen fit to appropriate funds for a census and classification of the plants of the United States. Hence, Shetler's comments on roses: "There are 20,000 cultivated varieties, and there are about 100 new ones being introduced every year. They are, from a botanical point of view, a nightmare to classify." He was concerned that the proposed national floral emblem was so ill defined from a scientific point of view.

Fellow scientist Gary S. Waggoner of the National Park Service then supported Shetler, reaffirming that from a botanical perspective the designation of a national floral emblem was premature. Waggoner suggested a review of possible candidates by professional botanists and horticulturists and wider public participation in the selection process.

At 3:55 P.M. the hearings were adjourned, as the subcommittee members had to get to the floor of the House to vote on Contra aid, a vote that may turn out to be as significant as the designation of a national floral symbol.

Appended to the hearing record is a history of the movement to designate a national flower, beginning with a vote for the goldenrod by the National Farmers Congress in 1889. The first bill before Congress in 1892 proposed the pansy. Subsequent proposals have favored mountain laurel, columbine, American dogwood, rose, corn tassel, marigold, grass, daffodil, Luther Burbank shasta daisy, blue morning glory, poinsettia, orange blossom, and sunflower.

Finally, on September 23, 1986, the House of Representatives passed H.J. Res. 385, identical to the Senate's S.J. Res. 159, and, following the President's signature, the rose became the national floral emblem. Sentiment and politics had taken precedence over botanical reasoning and taxonomic considerations.

The new law's impact on the country was less than overwhelming. My person-in-the-street interviews among people working on my floor revealed no one who knew that a national floral emblem had been selected. I could not detect much emotion about the subject one way or the other, yet the rose is clearly an emotional symbol. Random thoughts come to mind:

I am the rose of Sharon, and the lily of the valleys

Days of wine and roses

'Tis the last rose of summer

When you wore a tulip... and I wore a big red rose

I sometimes think that never blows so red
The Rose as where some buried Caesar bled

Everything's coming up roses

In the end I am left impressed by the caveats of the botanists who appeared before the committee. We have a national floral symbol that may be of the family Rosaceae or the genus Rosa or any one of the 200 species or 20,000 varieties. Just what is the national flower? How can I draw it or paint it or make a garland out of it to adorn the neck of the bald eagle?

Yes, Gertrude, life, as your words may indicate, *is* very complicated. However, I for one shall support our national floral emblem and display one in my lapel on the Fourth of July and other suitable occasions. I shall also endorse the efforts of plant scientists to classify all American flora.

44

Flights of Fancy

OVER THE YEARS, when our children were growing up, discussions at the dinner table often led to a postprandial examination of an encyclopedia to resolve some differences of opinion about matters of fact. Through the years we thus acquired two editions of the *Britannica,* one of the *World Book,* a few copies of the one-volume *Columbia,* and a *Book of Knowledge,* which came with me from my parents' home. It was thus an exercise in nostalgia, mixed with *déjà vu,* when the 17 members of the now-extended family were sitting around the New Year's table and the unlikely discussion of flying mammals arose.

I asserted, more surely than was justified by my solid knowledge, that all flying mammals were bats, members of the order Chiroptera. Exception was immediately noted: "What about flying foxes?" "How about flying squirrels?" and "Don't forget the flying wombat!" The matter clearly awaited encyclopedic authority.

On finally getting to the appropriate archival volumes, we found that

1. Flying foxes are large Old World bats belonging to the family Pteropodidae.

2. Flying squirrels are rodents that glide but do not truly fly.

3. The gliding mode of locomotion also applies to the flying lemur, or colugo, and the flying phalanger, an arboreal marsupial.

4. The flying wombat is unknown in my encyclopedias. On reflection, I believe it was the brand name of an audacious automobile in a 1940s comic strip.

For completeness in reviewing names that look as though they belong to flying mammals, I should point out that the Flying Tigers were a group of American airmen in the Burma theater of operations around 1941 and the *Flying Dutchman* is the name of a spectral ship doomed to sail forever. I hope this completes the catalogue of apparent flying mammals.

That left us free to find out about bats, and once having begun the game of "Reference Book," we were off and flying. Bats, as noted, form the mammalian order Chiroptera, which includes more than 900 species. They are the only mammals to have evolved true flight. Bats range in size from *Pteropus vampyrus,* with a 5-foot wingspan, to the tiny *Tylonycterus pachypus meyeri,* which stretches out to 6 inches but weighs a mere 1.5 gm as an adult. We will return to Tylonycterus later, for it is probably the smallest of all mammals.

Bats appear to have been around for 50 million years. Fossils of Icaronycteris (a prehistoric bat) found in Wyoming are at least that old. Like us, they trace their ancestry back to the insectivores. There are two suborders of bats that may have evolved independently from insectivores or may have had a common batlike ancestor. The two suborders sense their environment differently. One is visually oriented, and the other uses echolocation, or sonar. The value of the latter to a nocturnal animal is easy to understand.

Any discussion of bats seems to lead rapidly to vampires and thoughts about which came first: the human or chiropterous variety. Linguistically (*Oxford English Dictionary*), the word "vampire" derives from a very old Slavic term describing a preternatural being of a malignant nature supposed to seek nourishment or do harm by sucking the blood of sleeping persons. In this context the word seems to have entered English usage in the 1730s.

Bloodsucking bats are of three species confined to South and Central America. They were apparently first noted by Europeans in the 1498 landing of Columbus on Trinidad. All of these bats are of the family Desmodontidae, and reference to them as vampires appeared in English in 1774. The notion of bloodsucking humans appears to have been a part of European folklore long before true mammalian bloodsuckers were described.

The Desmodontidae show one of those biological adaptations that are often so fascinating. Their saliva contains an anticoagulant, so that the blood of their victims does not clog as they sip away. I have never read of an anticoagulant in connection with Dracula, but perhaps writers of this genre have not adequately perused writings on Chiroptera.

Now let us return to *Tylonycterus pachypus meyeri*. This wee creature caught my attention because the adult is reported as weighing 1.5 gm (the weight of a worn dime). I had thought the smallest mammal was the Etruscan shrew, which weighs in at 2 gm. However, I would guess that the wings of this bat are probably a third of its weight, so its body may be only about half the weight of the shrew, or 1 gm. Since the average human body weight is 60 kg, this tiny mammal is about 1/60,000 the weight of a human. If dimensions scale, a small gland that weighs 1 gm in the human body would tip the scales at 16 μg in Tylonycterus. This would still be 4×10^6 cells, so scaling should be possible.

The small bat loses body heat at about 40 times the rate of humans, so it comes as no surprise that it is a tropical animal. The species is found in the Philippines and gets its food from sugarcane, a rich and plentiful store of energy. I doubt that any smaller mammals exist, because the problem of heat loss would be overwhelming.

Well, that's as far as I can go with home reference books. But I think we all have a better understanding of flying mammals, and the next time a colleague pontificates that the smallest mammal is the Etruscan shrew, I will say, "But what about *Tylonycterus pachypus meyeri?*" With luck, this information will even make it into an edition of Trivial Pursuit, and if I know the answer, I will be able to have a grandchild say, "Awesome!" Right, Kris?

45

A Living Library

BACK IN THE EARLY DAYS of classification of organisms, zoos and botanical gardens kept living type specimens of each species for purposes of identifying and comparing unknowns. For backup, museums kept preserved specimens. As the number of known species began to grow dramatically, the task became increasingly difficult, and the discovery of microorganisms added a whole new dimension to the problem.

Preserving strains of microbes requires either continuous cultivation by serial passage or preparation of dried or frozen samples of those cultures that will survive these harsh procedures. At first, in the late 1800s, each investigator maintained his own strains, and other scientists obtained cell lines from the individuals who isolated them. There were serious problems with these procedures. Organisms evolve under continuous laboratory culture, and there is the ever-present danger of contamination.

In response to the need for standardized strains for research and commerce, scientists in various countries established type collections. In the United States at the American Museum of Natural History in New York, C.E.A. Winslow founded in 1913 a Bacteriological Collection and Bureau for the Distribution of Bacterial Cultures.

In 1925, a number of individuals and organizations, realizing the increasing scope and importance of maintaining microorganisms, established, in Chicago, a national organization, the American Type Culture Collection (ATCC). The original collection was restricted to bacteria and fungi. In 1937, when the ATCC moved to the Washington, D.C., area, the collection was expanded to include protozoa and algae. It was further expanded in 1949 to include animal viruses, and plant viruses were added in 1951. With developments in cell biology, new collections were needed. Starting in 1960, animal cell lines became part of the inventory, followed by the addition of rDNA vectors in 1981 and oncogenes in 1983.

I first learned of the ATCC in a bacteriology course in my first semester of graduate work and periodically ordered strains from it over the years. When my laboratory began the search for the smallest autonomous self-replicating cells, we started out by looking at each ATCC strain whose species name was *parvulus*. We eventually turned to mycoplasma, a form so new to laboratory cultivation that we still obtained most strains from the workers who had isolated them. As I worked on other projects over the years, I continued to order test strains from the ATCC and always had a special regard for those who made possible the systematic pursuit of microbial research. A few years of working at the National Bureau of Standards early in my professional days had developed my empathy for those who spend their careers making it possible for others to deal precisely with measurements and materials. The ATCC is very much in that tradition, playing a vital fiduciary role for biological research.

Recently, I had an opportunity to visit the ATCC; meet its director, R. E. Stevenson; and see firsthand the work of this vital organization. The view is worth recounting. Now located in Rock-ville, Md., the institution employs 125 scientists and technicians. Seen from Parklawn Drive, the ATCC headquarters is a modest, two-story brick building. The interior includes numerous labo-ratories, industrial-size freeze-drying facilities, and an impressive array of cryogenic storage devices. It feels a bit strange to be in a room where a million sealed vials of suspended life forms are stored. Everything about the ATCC is, as might be expected, neat and orderly. One gets the feeling that our national microbial repository is in good hands indeed.

The classic function of the ATCC is the distribution of reference strains, and this service has been offered since the start of the institution. In addition, the ATCC is a repository of strains in-volved in patentable processes, a function of considerable com-mercial importance. The organization also conducts research in areas relevant to its mission, and a workshop program teaches a variety of techniques using microorganisms.

The holdings of the ATCC collection are a guide to the sys-tematic nature of modern microbiology. There are 11,800 strains of bacteria and 515 types of bacteriophages. On hand are 17,800 varieties of fungi and mycoviruses. Among the obligate parasites of animals, the ATCC maintains cultures of 1,190 viruses, chla-mydiae, and rickettsiae. The remainder of the collection consists of 3,100 specimens, including protozoa, algae, antisera, plant viruses, plasmids, and oncogenes.

Living forms range in size from the tiniest mycoplasma 1/50,000 inch in diameter to the giant blue whale weighing 100 tons. The latter are faced with extinction, and we have no sure way to preserve them. The biodiversity problem thus weighs heavily on the minds of many biologists. At the microbial level, institutions such as the ATCC are a strong protection against species loss.

The suggestion has been made that we establish a library of tissue materials from endangered species in the hope that future technology will enable us to reestablish living organisms from those samples. That may be a bit futuristic, but the success of institutions such as the ATCC in maintaining its live collections encourages a rather heady view of the possibilities for living libraries.

In any case, the importance of maintaining archival repositories of living cells and molecules is now well recognized. That is our ultimate guarantee of a continuing source of defined living materials suitable for scientific exploration of the nature of life. The keepers of these collections deserve our thanks and appreciation.

46

Χάος,
Chaos chaos,
and Chaos

I SUPPOSE it is a fitting sign of the times that chaos has become, simultaneously, an important concept in science, the title of a widely read book, and a topic for cocktail party conversation. It is to the last of these subjects that this essay is addressed.

The English word "chaos" comes to us from the ancient Greek χάοσ. The earliest extant reference is from Hesiod, who lived in the eighth and seventh centuries B.C. In his *Theogony* (lines 114–116), he wrote:

ταῦτά μοι ἔσπετε Μοῦσαι, Ολύμπια δώματ' ἔχουσαι
ἐξ ἀρχῆς, καὶ ἔιπαθ', ὅτι πρῶτον γένετ' αὐτῶν.
Ἤτοι μὲν πρώτιστα Χάος γένετ'.
("O Muses who make Olympus home, tell me these things
From the beginning, and say who of these was the first to come to be.
Indeed the very first was Chaos.")

Thus, Chaos was the name of the most ancient of the Greek deities, the first member of the Pantheon. From very early times there was apparently an association between the first god, Chaos, and the unformed matter of the universe, chaos. The word "chaos" is first recorded in English in *Scala perfeccionis* by Walter Hylton in 1440. He writes of chaos as "a thycke derkeness."

The term "chaos" entered science in an unlikely way. The first recorded observation of an amoeba was by Rösel von Rosenhof in 1755. In the 10th edition of *Systema Naturae,* Linnaeus named this protozoan *Volvox chaos.* In a later edition, he changed the name to *Chaos protheus.* To anyone who has ever watched an amoeba under a microscope, the concept of chaos seems an appropriate basis for a genus or species name. In one of those endless disputes among taxonomists, there is some debate as to whether the original amoeba should be called *Volvox chaos* or *Chaos chaos.* Systematics aside, poetry would seem to dictate the latter name.

It is uncertain, however, which modern amoeba corresponds to the one sighted by von Rosenhof. A. A. Schaeffer in 1926 maintained that the giant amoeba *Pelomyxa carolinensis* is the original strain and should be designated *Chaos chaos.* Sufficient chaos exists in the classification of the naked Lobosa, however, that uncertainty persists about the taxonomic survival of *Chaos* or *chaos.*

The modern use of "chaos" has little to do with either mythology or protozoology but rather has to do with the physics of dynamical systems. Contemporary chaos, instead of being the father of the gods, is the child of computers. The computational mathematics on which it is based would have been so tedious and time consuming in the precomputer era that no one could have carried out the work needed to produce meaningful results. The existence of this kind of chaos was suspected as early as the 1890s by the French mathematician Henri Poincaré, but he had no way of demonstrating it with quantitative examples.

In classical physics or engineering, problems are approached theoretically by setting up the equations that govern the phenomena. These may be Newton's laws of motion, the equations of hydrodynamics, the laws of heat flow, or the equations of electromagnetism. For any given problem the appropriate dynamical equations are chosen. Dynamical equations show how given quantities change with time. The equations are solved relative to initial conditions. That is, at some time one must know the exact values of the variables from which one can compute the values at any time in the future, or so physicists thought some time ago. Indeed this was the basis of the mythical Laplacian demon, who from a knowledge of the present state of the universe could predict the state of the universe for all times in the future. Classical physics took on a Platonic character that allowed it to move from the ideal laws to a complete description of the shadows on the wall of the cave.

There was at least one problem with this approach. How precisely did the initial conditions need to be measured in order to solve for the future state of the systems? This question was largely ignored because it was widely believed that small differences in the initial state would lead to small differences in the final state. This assumption turned out to be true for those cases where the mathematics was tractable and numerical solutions could be obtained. In these cases great emphasis was put on linear approximations, which were the bread and butter of classical theoretical physics. The importance of the exact values of the initial conditions was also doubted because there was always experimental error, so that variables could only be known within a range of values.

The high-speed digital computer made it possible to use much more accurate numerical techniques in solving families of dynamical equations. When this was done, the result was surprising. Small differences in initial conditions could lead to very large differences, which deviated more and more as time went on. This effect was first clearly seen in the work of Edward N. Lorenz, a meteorologist at the Massachusetts Institute of Technology. In modeling the weather with a small number of differential equations, he found that the behavior was extraordinarily sensitive to small differences in initial conditions. A number of examples followed, and the science of chaos was born.

The result was shattering not only to the world of physics but also to the philosophy of knowledge. It means that the world can be perfectly deterministic and yet totally unpredictable. Given the lack of predictability, however, how do we know that the laws are deterministic? The chaotic behavior of a class of dynamical systems touches everything from long-range weather forecasting to the problem of free will. Our view of the physical universe, systematically developed since Galileo and Newton, must be reexamined not only at the atomic level of quantum mechanics but also at the macroscopic level in domains such as meteorology and geophysics.

Chaos represents a significant alteration in the way scientists look at the universe. The ideas are new and therefore difficult. They are well discussed in the popular book *Chaos: Making a New Science* by James Gleick. Perhaps Gleick oversells a bit, as most popular books are wont to do, but the subject is important enough for a bit of hyperbole to be forgivable. It is a book I can recommend with enthusiasm for anyone interested in these new developments.

It is sobering to realize that our most deeply held scientific views of the universe are so fragile that a new method of calculation can radically alter them. It is a helpful lesson in humility, for we scientists tend to be an arrogant lot. And the end is not yet in sight. Soon we shall have to deal with antichaos—those aspects of our world that cannot be chaotic. But that is another story, and my muse tires after all this deep and revolutionary thinking!

47

A Window in Time

THE SEARCH FOR LIFE'S ORIGINS is one of those quests that animate the spiritual side of our intellectual endeavors. Although we cannot tell at present exactly how life came to be, findings of the past 20 years in geology, biology, meteorology, and astrophysics enable us to pin down with surprising precision the date the ancestors of current living forms made their appearance on earth.

Knowledge of past life has come to us mainly through fossils, the hardened remains of plants and animals or the mineralized traces of those remains. Almost all of the fossils that fill our great museums and illuminate our biological past, however, are quite recent arrivals on a planet whose beginnings go back to somewhere between 4.5 and 4.6 billion years ago. The great age of fossils—the Phanerozoic Age—covers the most recent 570 million years, or only about one-eighth of the total time that the planet has existed.

The explanation for the relatively recent age of most fossils is twofold. The first cause is geological. The older the fossil-bearing rocks are, the longer the time they have had to weather and metamorphose (change by natural processes caused by temperature and pressure), and the harder it is to discern remains of once-living organisms. The second reason is biological. Most fossils come from teeth, bones, shells, wood, and other hard parts of plants and animals, and the organisms that lived before the beginning of the Cambrian Period, some 580 million years ago, had not yet evolved to develop these structures. Thus, up until a few years ago, although some fossils from late Precambrian times had been found, there seemed little hope of finding fossils from the early stages of life on earth.

Science often proceeds in unexpected ways by making connections between what at first might seem to be unrelated findings. Thus, there appeared to be no great theoretical significance to the discovery reported in 1825 of strange layered limestone structures around Saratoga Springs, N.Y. The formations, from the Precambrian Era, can be seen there today at the Petrified Sea Gardens. The next published report on these materials was not made until 1883, when James Hall postulated that they were of biologic origin and coined the name *Cryptozoon proliferum* for the species that had given rise to the fossils.

In 1908, Ernst Kalkowsky, working in northern Germany, found samples of similarly layered, or laminated, rocks. He estimated that these structures were from the Triassic Period, approximately 200 million years ago, and gave them the descriptive name "stromalith" from the Greek *stroma* ("mat") and *lithos* ("rock"). These rock-mat structures, widely found around the world, have since been renamed stromatolites, and they probably include the oldest fossils known on earth.

In 1914, Charles Doolittle Walcott came up with the idea that stromatolites resulted from the activities of a group of primitive blue-green algae. His idea followed from a study of present-day algae that grow in freshwater streams and lakes and in precipitate layers of calcium carbonate on mats of growing cells. He proposed a similar origin for stromatolites, which had an analogous appearance. One puzzling aspect of Walcott's idea was that the fossil stromatolites were associated with what had been saltwater environments, whereas the systems he studied were in freshwater surroundings.

Living saltwater stromatolites were subsequently discovered in 1933 by Maurice Black on Andros Island in the Bahamas. He found coastal areas where layered algal structures grew. These closely resembled the fossil stromatolites. The oceanic structures form somewhat differently from the freshwater ones. The former begin with the growth of a mat of algae that traps a layer of sedimentary particles on the algal filaments. The next layer grows on top of the sediments, followed by another sedimentary layer, which leads to the repetitive lamellar structures characteristic of stromatolites. Contemporary forms have been found in abundance in the intertidal flats of Florida and Western Australia. The growing mats are formed by a collection of different organisms rather than by a single species, as was originally assumed.

As attempts began to establish the dates of fossil stromatolites and the great age of a number of them became apparent, geologists reasoned that these objects were probably the oldest remains of life on earth. During the 1950s the use of isotopic dating techniques confirmed that a number of these specimens were indeed from the distant past. For the oldest fossils, the samarium-to-neodymium ratios and the uranium-to-lead ratios provide the best available dating.

The three oldest groups of stromatolites that have been dated are the Warawoona from Australia (dated as 3.56 billion years old), the Onverwacht from Africa (dated as 3.54 billion years old), and the Insuzi from Africa (dated as 3.09 billion years old). The most exciting of these from a biological point of view are the Warawoona materials.

These materials show a $^{13}C - ^{12}C$ isotope ratio that is characteristic of photosynthetic fixation. On cutting these materials in thin section and examining them by various forms of microscopy, one sees objects that are of a size and shape consistent with modern-day cyanobacteria (blue-green algae). Thus, it appears that by 3.56 billion years ago there was a community of organisms that showed

- Wall formation, necessary for the shapes of the objects seen.

- Photosynthesis, both from the carbon isotope ratios and from the fact that modern stromatolites are formed from photosynthetic organisms.

- Some kind of motility and phototaxis, deduced from the fact that organisms from the layer covered with sediment must move up toward the sunlight and start growing out again.

- Mat formation, a necessary property for stromatolite growth.

Thus, there is good evidence for a sophisticated ecosystem at Warawoona.

The oldest known rocks on the earth (there are older lunar rocks) are the 3.8-billion-year-old samples from the Isua formation in Greenland. These show $^{13}C - ^{12}C$ ratios characteristic of photosynthesis and objects that some paleontologists suggest are cellular remains, although this is disputed by others. Complex unicellular communities existed 3.56 billion years ago, and some evidence suggests life predating 3.8 billion years ago.

So what is the earliest time that present-day life forms could have existed? Here the evidence comes from astrophysics, lunar geology, geophysics, and meterology. It now seems clear that the earth was subjected to intense meteoric bombardment up to about 3.8 to 3.9 billion years ago. Some of these meteors were so large that they produced enough heat to evaporate the entire ocean. Such an event would have sterilized the earth's surface, destroyed any primordial life, and provided an event horizon for life as we know it. A conservative estimate is that the very large impact came somewhere around 4 billion years ago, which was about the time the planetary environment was coming into a state suitable for living cells to form.

Putting all of these factors together suggests that the window in time for life to form was between 3.7 and 4.1 billion years ago or, more likely, between 3.8 and 4.0 billion years ago. This may not seem to be a very precise estimate, but it does indicate that when conditions were right, life formed rather quickly. This novel finding provides us with some new and exciting ways of thinking about the origins of life. The next few years may well provide striking insights into the question of how life came to be on our planet. Knowing when is a good first step on the way to knowing how.

48

Planetary Protection

IN A MODEST OFFICE in NASA headquarters in Washington, D.C., sits a young man who bears the awesome title "Planetary Protection Officer." He is, I believe, the fourth or fifth person to carry this title: I have personally known at least three of his predecessors. The job of this official concerns itself with controlling interplanetary biological contamination accompanying missions from Earth to other planets and the possible return of extraterrestrial samples. I am in favor of someone's having this responsibility, and I sleep easier knowing that the matter of planetary protection is part of the federal agenda. Since missions to Mars are planned for the first quarter of the twenty-first century, the problem of planetary protection is beginning to warm up once more.

I was involved in the last great dispute over planetary protection, and retelling the story may prove to be a useful exercise. The time was 1969, a few months before the launch of the first manned lunar module, when plans were being made for three humans to go to the moon and return to Earth with a collection of moon rocks. It was an exciting time for lunar scientists and for a nation and a world excited by the idea of men on the moon.

The planetary biology subcommittee, on which I served, was meeting to give its final imprimatur to the biological aspects of the mission. The plan called for the return capsule to splash down in the Pacific Ocean and be picked up by a navy vessel. The entire capsule was then to be bolted to a trailer, the interior of which was to constitute an isolation chamber for astronauts and samples. The trailer, with capsule, was then to be taken to Houston, where a special laboratory—including housing for the astronauts during a several-week quarantine period—had been built.

In Houston, the trailer was to be connected to the lunar return laboratory so that the astronauts and samples could be kept in isolation and the samples could be worked on through a biological isolation barrier. The whole operation was exquisitely planned to minimize any biological hazard to planet Earth from returned samples or returned space explorers.

After we had reviewed the detailed procedures, a representative from NASA arrived on the scene to announce that a glitch had developed in the isolation scheme. The davit on the naval vessel was not rated to handle the fully loaded weight of the returned capsule. Therefore, the astronauts would climb out of the capsule into a dinghy and be taken by helicopter to the shipboard trailer, where they would go into isolation.

The committee members looked at each other in disbelief. The new procedure was a complete break in quarantine. The Earth's environment was to be exposed to any contaminants carried on the astronauts, and the men were to be subject to terrestrial contamination.

A heated discussion ensued. The purists argued for total protection of the planet and delay of the mission until an appropriate lifting device could be obtained. The pragmatists argued that the mission had been planned for a long time, and enormous political and public relations fallout would accompany any delay. The more whimsical among us argued that the very expensive and cumbersome isolation and quarantine procedures were folly since cross-contamination would inevitably occur after splashdown.

The meeting ended without any consensus. Although a majority of the committee were willing to proceed with the mission, there was a great deal of residual discontent. Somehow, all previous discussions of contamination were being reduced to the theater of the absurd.

Lurking behind the discussions, there was indeed a consensus: Everyone was convinced that the probability of viable organisms on the moon was vanishingly small. The whole quarantine procedure was developed from the perspective of being ultraconservative with the unknown. But if this was a valid strategy, what difference would a few months make in delaying the mission to solve the davit problem? And if it was not a valid strategy, why all the precautions and heavy expenditures at the Lunar Return Laboratory? The answer was not long in coming.

Members of the subcommittee began to receive phone calls from various NASA officials—including the Planetary Protection Officer, who was most distressed by the entire turn of events. The thrust of the phone calls was to alert the committee to the fact that this was a very tricky business politically. Any public opposition to the mission would have to be justified on the basis of the hazards of contamination being greater than the political hazards. As one subcommittee member was told: "There is only one person in the world who can delay this mission now, and his name is Richard Nixon."

Well, the launch took place on time, and it was clearly one of the most successful explorations of all time. The returned lunar materials showed absolutely no signs of life. Studies of the materials have been of greatest value in planetology, geochemistry, and geophysics. From a public relations point of view, the project was the utmost success.

Now there are tentative plans for an unmanned Mars mission that will return Martian samples to the Earth for study. Once again, the Planetary Protection Officer has to face the possibility of biotic contamination from another planet. With Martian samples, the problem is more difficult than with lunar samples— Mars is much more Earth-like than the moon, and there is evidence that water once existed on Mars. For this trip we have considerably more lead time. We also must be concerned with contaminating Mars with earthly material, which has probably happened already in any case.

In a sense, we are now better prepared than was the first Planetary Protection Officer. From past experience, we better understand the issues, whether those issues be scientific or political.

The more I have thought about the reverse-contamination problem, the less concerned I have become about the hazard to terrestrial life. The pathogens we know have co-evolved with their hosts and use a variety of biochemical pathways shared by all earthly life. If the pathways of biochemistry and molecular biology are truly universal, then extraterrestrial pathogens (or EPs, with apologies to Steven Spielberg) might survive a short time on Earth, but it is hard to believe that they could out-compete organisms that have been evolving for 3.8 billion years or more on this planet. Surely, it will not be possible to provide complete isolation for returned matter forever. As we learned in the lunar case, breakdowns in isolation do occur.

Total security is impossible. If we are to explore the universe, some risk is always present, and surprises will occur.

49

Rocks

SOMETIMES, when musing about seemingly insoluble problems, the mind takes to wandering along rather whimsical ways. So it was the other day when the tangible I was wandering down Front Street in Lahaina, Hawaii, thinking about the changing global ecosystem. I had been working on a piece about balancing species endangerment against economic concerns. That is a most difficult issue, truly deserving of Solomonic wisdom—which we, alas, can only aspire to.

In an unexplained thought, doubtless brought on by direct noonday sun on the forehead, I began to put myself in the position of a thinking rock about 3.85 billion years ago. I would have been distressed about what was happening to my environment. An ugly stuff referred to by some rocks as "life" is beginning to show up all over. The beautiful crystal-clear ocean waters are beginning to cloud up as this material accretes in very unrocklike ways. Clean wet seashore stones are being covered by an ugly, green-gray slimy goo. Rock environmentalists among us are truly upset.

Not only is the sea affected but the atmosphere itself is being changed in a threatening manner. Traces of highly toxic oxygen are being released into our atmosphere. This insidious material takes the beautiful bodies of ferrous iron and covers them with a nasty, flaky red iron oxide, or rust. Oxygen is a potential threat to all the world's great ferrous iron deposits. The rapidly growing menace of life is also taking carbon dioxide out of the atmosphere, thus upsetting the long-standing balance of nature. As the carbon dioxide levels drop, beautiful marine crystals of calcium carbonate are being dissolved.

Not since the great meteor showers ceased a hundred million years or so ago has there been anything so dangerous to environmental quality. This is truly a catastrophic situation.

Some speculate that if more oxygen is released, then even more reactive ozone will form in the upper atmosphere, and this will rob us of the purifying and revivifying shortwave ultraviolet light, which cleans contaminated surfaces. All of rockdom is horrified by what is happening. No way is known to stop this menace that seems to be increasing at an exponential rate. Somehow, we must return to the good old days, the world as it was before this noxious life somehow came to be. Our clean, crystalline, inorganic world is being polluted by this indescribably gross, amorphous material, which defies description of its ugliness.

Of course, my lithic protestations would have come to nought. Once it had started, there was no stopping this highly adaptive life that has occupied and changed every nook and cranny in the world. For almost 4 billion years, life has flourished and radically changed the atmosphere, the hydrosphere, and even the lithosphere. The record shows that change has been faster or slower from time to time. But change has always characterized the world.

A few billion years later, my thinking rock would have been no more, weathered into oblivion or recycled by subduction or volcanism into some unrecognizable form. At that later time, another change on the planet was introduced, an alteration as profound as the earlier transition to life. Along the primate line something was coming into being as radical and as different as early life itself, and ultimately capable of changing the planet just as radically.

One branch of the primates evolved into the hominid line, and there followed Australopithecus, *Homo habilis, Homo erectus,* and finally *Homo sapiens.* "Thought" appeared on the surface of the earth. Thought is as powerful and as capable of changing the planet as is life. It is much more difficult to conceptualize the emergence of thought, for we are dealing with something less tangible than the gooey ooze on the rocks. Nevertheless, the world has been radically altered.

For the first couple of hundred thousand years, *Homo sapiens* existed as a hunter-gatherer, and the full potential of thought began to emerge only slowly. For the past 10,000 years, thought has gone explosive—and with it there have been changes in the environment of unprecedented dimensions. The agricultural revolution led to a massive conversion of forests, plains, and savannas to fields and gardens. From a conservationist's point of view, this is a catastrophe, with massive loss of species and change of habitats. To René Dubos in *The Wooing of Earth,* it is the humanization of the planet, the change of the environment to suit us rather than vice versa, which is the usual biological imperative.

It is not possible to hold back the consequences of thought any more than it was possible for rocks to restrain the consequences of life. What then is the problem? There is a continuity in the evolution of the universe. The coming of life did not do away with the geological nature of the planet; life adapted to it and altered it. The coming of thought did not abrogate the biological nature of *Homo sapiens;* we are still very much animals. As animals, the increase in our numbers is part of our very nature. Thus, we procreate and at the same time develop ways of modifying the world to accommodate our ever-increasing numbers and our ever-increasing appetites for manufactured goods.

We may lament the environmental effects of the age of human thought, but we cannot eliminate them any more than the rocks could have blocked the sequelae of the coming age of life. Change is a property of the evolving globe. The emergence of the noetic element has discontinuously accelerated the rate of that change.

There is, of course, something very different about the evolution of the mind as compared with all other biotic transitions.

A reflective element is introduced: We can think about the consequences of our actions. As a result, a radically new feed back
element is introduced in the unfolding of global history. What
we cannot do is to continue life on the planet in its prehuman
state of evolution. We can have some choice in how to change
the planet, but we cannot opt to leave it unchanged. The prehuman earth is considered by many a Garden of Eden, a mythical
paradise. It was, in fact, a highly competitive Darwinian world
in which our distant ancestors struggled for survival. We hom-
inids have been so successful in winning that battle that we are
now in danger of altering the planet to our own detriment. That
prospect requires intelligent management. We, however, cannot
go back.